T0315015

EXPOSURE ASSESSMENT AND SAFETY CONSIDERATIONS FOR WORKING WITH ENGINEERED NANOPARTICLES

EXPOSURE ASSESSMENT AND SAFETY CONSIDERATIONS FOR WORKING WITH ENGINEERED NANOPARTICLES

MICHAEL J. ELLENBECKER, Sc.D., CIH
Professor Emeritus of Occupational and Environmental Hygiene
Director, Massachusetts Toxics use Reduction Institute
Department of Work Environment
College of Health Sciences
University of Massachusetts Lowell
One University Avenue
Lowell, MA, USA

CANDACE SU-JUNG TSAI, Sc.D.
Assistant Professor of Occupational/Environmental Health and Hygiene
School of Health Sciences
College of Health and Human Sciences
Purdue University
Delon and Elizabeth Hampton Hall of Civil Engineering
550 Stadium Mall Drive
West Lafayette, IN, USA

Library of Congress Cataloging-in-Publication Data:

Ellenbecker, Michael J.
    Exposure assessment and safety considerations for working with engineered nanoparticles / Michael J. Ellenbecker, Sc.D., CIH, Professor Emeritus of Occupational and Environmental Hygiene, Director, Massachusetts Toxics use Reduction Institute, Department of Work Environment, College of Health Sciences, University of Massachusetts Lowell, Su-Jung (Candace) Tsai, Sc.D., Assistant Professor of Occupational/Environmental Health and Hygiene, School of Health Sciences, College of Health and Human Sciences, Purdue University.
        pages   cm
    Includes bibliographical references and index.
    ISBN 978-0-470-46706-0 (cloth)
1. Nanotechnology–Safety measures.   2. Nanostructured materials industry–Employees–Health and hygiene.   3. Nanotechnology–Health aspects.
4. Nanoparticles–Toxicology.   5. Industrial hygiene.   I. Tsai, Su-Jung.   II. Title.
    T174.7.E454 2015
    363.17′9–dc23
                                                                                    2014029773

Cover image credit:
Image courtesy of Argonne National Laboratories
Image title: Fireworks over Night Sky
Image was taken by Vilas Pol and Natalie Fitzgerald
Description of image:
A dramatic fireworks display over a night sky is demonstrated. Golden yellow lights pournward along with colorful and radiant clusters of pink, purple and orange. The viewer's perspective is that of looking up at the fireworks coming down to the earth in a shower of light. Originally, it's a scanning electron micrograph of photo-luminescentlanthanum hydroxycarbonate superstructure, prepared in critical autogenic reactions. The U.S. Department of Energy's Argonne National Laboratory, 2011

1   2015

Dedication

I (ME) dedicate this book to my wife, Marlene Goldman, for her understanding throughout the long process of research and writing leading to its publication. I greatly appreciate her support. I also dedicate it to our daughters, Anne and Heidi, and our five grandchildren, in the hope that the safe application of nanotechnology will bring them a better world.

I (CT) dedicate this book to my husband, Chunyuan Chen, and children, Xenia and Steven, for their understanding and encouragement while I spent the time doing the research followed by the writing. I appreciate my parents' support of my career. Being in the forefront conducting research on this important topic, I hope that nanotechnology will bring a bright future to the next generation with a safe environment for everyone.

CONTENTS

PREFACE

The field of nanotechnology is expanding rapidly, with new discoveries being announced almost daily in fields ranging from the fairly mundane (e.g., advanced composite materials) to the exotic (e.g., targeted anticancer drug delivery). As with any new technology, however, excitement over the likely benefits to society must be balanced with attention to any potential adverse effects that may arise. For nanotechnology, there is growing evidence that at least some of the new engineered nanomaterials may have adverse health and safety aspects that must be addressed in order to ensure that they are used in a safe and effective manner.

Hence this book. We the authors, occupational hygienists, have been performing research into the occupational and environmental health and safety aspects of nanotechnology since 2003. Here we have attempted to present our view of the current state of this rapidly advancing field. The primary audience for the book is health and safety professionals who find that they have to develop a working knowledge of this new field. Given our backgrounds, the book's primary focus is on occupational exposures, and we have attempted to give attention to environmental exposures. It is our hope that the reader will be able to use the information from this book to apply the occupational hygiene model of anticipation, recognition, evaluation, and control to nanotechnology development.

We would like to acknowledge the contributions of Rick Reibstein of the Massachusetts Office of Technical Assistance and Technology, who provided valuable information and insight to Chapter 11. We also acknowledge the support received from the National Institute for Occupational Safety and Health (NIOSH) and the National Science Foundation (NSF) funded Center for High-rate Nanomanufacturing, which supported much of our research as reported in the book's case studies.

Michael J. Ellenbecker
Candace Su-Jung Tsai

1

INTRODUCTION

1.1 WHY A BOOK ON NANOTECHNOLOGY HEALTH AND SAFETY?

Asbestos, once hailed as a "miracle" material for its insulating properties, has been an occupational and environmental health disaster. Many thousands of people, mostly workers but also members of the general population, have developed serious illnesses, including asbestosis, lung and colon cancer, and mesothelioma, and many have died as a result of their exposure. The reader may ask, "why start a book on nanoparticle health and safety with a discussion of asbestos, which most definitely is not a nanoparticle?" The answer leads us to our purpose in writing this book, at this time.

The authors have been to many nanoparticle health and safety meetings over the past several years. A constant theme at those meetings, both in the formal presentations and also in the informal discussions among the scientists in attendance, is "we have to prevent the next asbestos." Starting in the 1920s, exposure to asbestos from its mining, milling, and incorporation into products such as textiles was associated with severe lung disease that came to be called asbestosis; other hazards from asbestos exposure, such as lung cancer and mesothelioma, were not discovered until years later. In addition, the risks to workers using products containing asbestos and to individuals in the general population took some time to be appreciated.

Today, engineered nanoparticles represent a miracle new material (actually, a range of materials, as discussed later), just as asbestos was a miracle new material

Exposure Assessment and Safety Considerations for Working with Engineered Nanoparticles,
First Edition. Michael J. Ellenbecker and Candace Su-Jung Tsai.

early in the last century. And, as with asbestos, there are early indications that there may be adverse health effects associated with at least some of these new materials; in fact, carbon nanotubes may have similar health effects as asbestos (see Section 5.4). The extent of the risk to workers and the general public is not known at this time. The answer to the question posed above leads to another question, that is, have we learned our lessons from asbestos and other similar occupational and environmental health disasters, so that we can develop the exciting new field of nanotechnology while protecting the health of workers and the general population, and prevent any adverse effects to the environment?

We believe that the answer to this question is "yes." The nanotechnology industry is still in its infancy, meaning that proactive steps can be taken to further its development in a safe, sustainable manner. Andrew Maynard and colleagues summarized the risks and opportunities in their 2006 *Nature* article as follows (Maynard et al., 2006):

> The spectre of possible harm—whether real or imagined—is threatening to slow the development of nanotechnology unless sound, independent and authoritative information is developed on what the risks are, and how to avoid them. In what may be unprecedented pre-emptive action in the face of a new technology, governments, industries and research organizations around the world are beginning to address how the benefits of emerging nanotechnologies can be realized while minimizing potential risks.

This book has been written in an attempt to contribute to the minimizing of the potential risks of nanotechnology. In occupational and environmental health, we have a very simple model that guides our work, that is, exposure to a material or physical agent may lead to an adverse health effect in the exposed population. Although we have included a brief review of the current state of nanoparticle toxicology in Chapter 5, in order to put the need for exposure assessment and control in their proper context, this is a book about the exposure side of our model. In considering exposure, the two most important aspects are to *evaluate* the magnitude of the exposure and, in those cases where the exposure is judged to be excessive, take steps to *control* the exposure. These are the two major topics covered in this book.

It is important to emphasize that this book pays relatively little attention to the *judgment* step just mentioned. In most cases, environmental health professionals make the decision as to whether a measured exposure is excessive by comparison to standards, such as published occupational exposure limits. These standards, in turn, are established based on results of toxicology and epidemiology studies that quantify the risk of exposure for a certain material. The difficulty with engineered nanomaterials, as discussed in Chapter 5, is that at this time there is insufficient information to set such standards. The consensus among occupational and environmental health scientists studying engineered nanoparticles is that until sufficient toxicology and epidemiology information is available for any given material, the precautionary approach must be followed in order to minimize the risk to workers, the general public, and the environment. This concept is further discussed in Section 1.4.

In the occupational environment, exposure assessment and control are the purview of industrial hygiene or, more widely used today, occupational hygiene; regarding the general environment, the equivalent field might be called "environmental hygiene," although this term is not widely used. In any case, the subject of this book is the current state of knowledge concerning the occupational and environmental hygiene aspects of nanotechnology. It is fair to say, however, that most of the focus in on occupational hygiene, and there are two reasons for this. First is the fact that the authors are occupational hygienists, so most of our experience and expertise, such as it is, falls within this field. The second reason is perhaps more important, which is that in a new and growing industry such as nanotechnology, most of the significant exposures will be to those workers who are doing research with and manufacturing nanomaterials. Consequently, most of the concerns and attention of the research community to date has been focused on nanotechnology workers, rather than the general public. As nanotechnology-enabled products become more widely used, we can expect more of the focus to shift to their environmental impact.

1.2   SOME SCENARIOS

Some scenarios may help to put the need for nanoparticle health and safety in its proper context. These scenarios are all fictional but based more or less on real situations now being encountered in the nanotechnology field.

- A small company specializes in the manufacture of relatively small quantities of high-quality powders in the micrometer size range for specialized niche markets. The company is too small to employ a person with health and safety training. The company president wishes to modify their production equipment to produce powders in the nanometer size range but has read of general concerns about the safety of nanoparticles. He asks his production manager to find out whether manufacturing nanopowders will present any new hazards to their workers.

- A university laboratory is conducting research on the use of carbon nanotubes (CNTs) as a new form of digital storage device. As part of this research, the laboratory technician must transfer small quantities of dry bulk CNTs from a 2 L jar to a beaker, weigh them on a balance, and disperse them in a solvent. These steps are now done on a laboratory bench, and the technician is concerned that she may be breathing in some CNTs that are being released from the powder.

- A plastics manufacturing company has recently begun pilot-scale testing of a new composite material, consisting of polyester reinforced with 2% by mass of alumina nanoparticles. The composite is produced in a twin-screw extruder. The polymer pellets are dumped into one hopper and the nanoalumina in another; they are fed by gravity into the throat of the extruder, where the pellets are melted by high temperature and pressure and mixed with the nanoalumina to form the nanocomposite. The extruder operator has noticed some of the alumina powder on various work surfaces around the hopper and is concerned that this may not be completely safe.

- A company uses the nanocomposite material produced above to make tennis racket frames. The pellets are fed into an injection molding machine which produces the tennis racket shape. However, extraneous material has to be cut off the frame with a saw and the frame must be sanded smooth. The plant occupational hygienist is concerned that these operations may release nanoparticles. In addition, the tennis racket manufacturer labels the product "contains nanomaterials for added strength" and is receiving anxious questions from consumers on their web site.
- An occupational hygienist working in the health and safety office at a large research university conducted a university-wide survey which found that more than 50 laboratories across the university claimed to be doing research with some type of nanoparticle. Many of the respondents stated that they did not know what practices should be followed when working with nanoparticles; the occupational hygienist decided that the university needed a policy on good work practices but he did not know how he would gather the information to develop such a policy.
- A research laboratory recently purchased a chemical vapor deposition furnace for making CNTs. While the furnace is a closed system during CNT production, at the end of a run, it must be opened and the CNT material must be scraped from the furnace walls into a drum. This process creates visible dust, which has caused the operator some anxiety.
- A research laboratory was working with quantum dots suspended in a solvent. During normal operations, the laboratory director was fairly confident that no nanoparticle exposure was occurring, since the particles were in liquid suspension. One day, however, a technician dropped a beaker containing the nanoparticle suspension, and the beaker shattered when it hit the floor. Since the laboratory had no emergency response plan in place, the laboratory director was unsure what steps to take to clean up the spill. She decided to evacuate all personnel from the lab, waited until the solvent had completely evaporated, swept up the broken glass and placed it in the lab's broken glass receptacle. The workers were then allowed back in the lab.

These scenarios share certain characteristics, that is,

- Uncertainty as to the adverse health effects, if any, associated with using a nanomaterial;
- Uncertainty as to the level of worker exposure, if any, from using a nanomaterial; and
- Uncertainty as to what steps (control methods) should be taken to ensure that worker exposures are controlled to an acceptable level.

This text does not address the first point above but is meant to address the other two. We hope to provide the reader with the latest information concerning techniques to evaluate and control occupational and environmental exposures to nanoparticles.

1.3 ORGANIZATION OF THE MATERIAL

The remainder of the book generally follows the basic occupational hygiene model for addressing a possible worker exposure, which is the four-step process of anticipation, recognition, evaluation, and control. We start in Chapter 2 with a presentation on the terms used in this field—the definition of a nanoparticle, the different types of nanoparticles, and so on. In the anticipation phase, we attempt to identify possible problems of concern. Chapter 3 discusses the unique properties of nanoparticles that may make them more toxic than larger particles of the same material, and Chapter 4 describes the various pathways by which a nanoparticle might enter the body. We then move to the recognition phase in Chapter 5 with an overview of the current knowledge on engineered nanoparticle toxicity, followed by a discussion of possible sources of exposure in Chapter 6.

We then move to the main topics of this book, evaluation, and control. Chapter 7 reviews the current state of instruments and methods available to evaluate exposures to engineered nanoparticles, and Chapter 8 reviews in detail how those tools are used to perform an engineered nanoparticle exposure characterization. Included in Chapter 8 are three detailed case studies arising from work in our laboratory that illustrate the practical application of the exposure assessment methodology. Chapters 9 and 10 are concerned with solutions—methods to control occupational and environmental exposures that an exposure assessment found to be problematic. Again, the basic techniques are supplemented by case studies from our research.

We end the book by considering important policy issues facing the nanotechnology health and safety community. Chapter 11 discusses the current state of health and safety regulations, both in the United States and internationally, and Chapter 12 concludes with our thoughts and recommendations for needed next steps if we are to continue to make progress toward our goal of preventing occupational injuries and disease related to nanotechnology.

1.4 OUR APPROACH TO NANOPARTICLE HEALTH AND SAFETY

We will return to this topic at the end of Chapter 12, but we believe that it is important for the reader to understand the basic philosophy behind our approach to nanoparticle health and safety, since it informs all of the topics covered throughout the text. At the time of this writing in 2014, much is not known about the toxicology of engineered nanoparticles, but enough is known to certainly raise concerns. Likewise, we know that workers and the general public have the potential to be exposed to engineered nanoparticles. Therefore, we believe that a precautionary approach must be followed—specifically, we believe that until toxicology and/or epidemiology tell us differently, all reasonable steps should be taken to minimize exposures to engineered nanoparticles. In order to accomplish this, exposures must first be measured and, where exposures are found to be measurably above background levels, steps must be taken to reduce those exposures.

This approach is not unique to us—indeed, many researchers and government agencies responsible for nanotechnology health and safety have taken similar positions. The precautionary approach relies on the concept of the "precautionary principle," first developed in Europe and adopted into the 1987 Ministerial Declaration on the protection of the North Sea (Anonymous, 1987), which stated, regarding contamination of the North Sea by chemicals, that "…a precautionary approach is necessary which will require action to control inputs of such substances even before a causal link has been established by absolutely clear scientific evidence." The precautionary principle was affirmed at the United Nations Conference on Environment and Development, held in Rio de Janeiro in June 1992; Principle 15 of the Rio Declaration on Environment and Development (UNEP, 1992) states:

> In order to protect the environment, the precautionary approach shall be widely applied by States according to their capabilities. Where there are threats of serious or irreversible damage, lack of full scientific certainty shall not be used as a reason for postponing cost-effective measures to prevent environmental degradation.

The precautionary principle has been applied by others when addressing engineered nanoparticle health and safety. For example, the US National Institute for Occupational Safety and Health (NIOSH) has written (NIOSH, 2009):

> **Until further information on the possible health risks and extent of occupational exposure to nanomaterials becomes available, interim protective measures should be developed and implemented.** These measures should focus on the development of engineering controls and safe working practices tailored to the specific processes and materials where workers might be exposed *[emphasis in original]*.

Such recommendations of caution are not limited to US sources. For example, the British Standards Institute issued the following recommendation (BSI, 2007):

> This Published Document recognizes that there is considerable uncertainty about many aspects of effective risk assessment of nanomaterials, including the hazardous potential of many types of nanoparticles and the levels below which individuals might be exposed with minimal likelihood of adverse health effects. The guide therefore recommends a cautious strategy for handling and disposing of nanomaterials.

ASTM International, a widely recognized consensus standard-setting organization, recommends the following (ASTM, 2007):

> Exposure control guidance in this Guide is premised on the principle (established in this guide) that, as a cautionary measure, occupational exposures to UNP [unbound, engineered nanoscale particles] should be minimized to levels that are as low as is reasonably practicable. This principle does not refer to a specific numerical guideline, but to a management objective, adopted on a cautionary basis, to guide the user when

(*a*) assessing the site-specific potential for such exposures; (*b*) establishing and implementing procedures to minimize such exposures; (*c*) designing facilities and manufacturing processes; and (*d*) providing resources to achieve the objective.

Circling back to where we began, with asbestos, Schulte and colleagues recently published an article (Schulte et al., 2012) that reviewed what is known of the hazards of CNTs and recommended a program of action. They conclude:

> In the evolution of human civilizations, learning from the history and not repeating it has been a key guiding principle. Society can learn from how asbestos was inappropriately considered and not make the same mistake with CNTs. It is possible to safely realize the benefits of CNTs, but it will require rigorous and timely actions. The time to act is now.

We firmly believe that, not just for CNTs but for engineered nanoparticles in general, we must act now to ensure that this exciting new industry moves forward in a way that is protective of human health. It is with this conviction that this book is written.

REFERENCES

Anonymous. Ministerial declaration on the protection of the North Sea. Environ Conserv 1987;14:357–361.

American Society for Testing and Materials [ASTM]. *Standard Guide for Handling Unbound Engineered Nanoscale Particles in Occupational Settings*. Philadelphia (PA): ASTM; 2007.

British Standards Institute [BSI]. *Nanotechnologies—Part 2: Guide to Safe Handling and Disposal of Manufactured Nanomaterials*. London: BSI; 2007.

Maynard A, Aitken RJ, Butz T, Colvin V, Donaldson K, Oberdörster G, Philbert MA, Ryan J, Seaton A, Stone V, Tinkle S, Tran L, Walker N, Warheit DB. Safe handling of nanotechnology. Nature 2006;444:267–269.

Schulte PA, Kuempel ED, Zumwalde RD, Geraci CL, Schubauer-Berigan MK, Castranova V, Laura Hodson L, Murashov V, Matthew M, Dahm MM, Ellenbecker MJ. Focused action to protect carbon nanotube workers. Am J Ind Med 2012;55:395–411. DOI: 10.1002/ajim.22028.

United Nations Environment Program [UNEP]. Rio declaration on environment and development E.73.II.A.14. Stockholm: UNEP; 1992.

U.S. National Institute for Occupational Safety and Health [NIOSH]. *Approaches to Safe Technology—Managing the Health and Safety Concerns Associated with Engineered Nanomaterials*. DHHS (NIOSH) Publication No. 2009-125. Washington (DC): NIOSH; 2009.

2

WHAT IS A NANOPARTICLE?

Although there is a strong current interest in nanoparticles because of the exciting possibilities for new products and whole industries utilizing nanotechnology, nanoparticles are not new. Nanoparticles occur naturally in the environment and have had impacts (mostly adverse) throughout human history. The most common *natural* nanoparticles are those produced by forest fires; nanoparticles produced in volcanic eruptions, although occurring much less frequently than in forest fires, are much more dangerous to humans and the environment. Another commonly encountered natural nanoparticle found in coastal areas is the sea salt aerosol produced by the oceans due to wave action.

The dawn of the Industrial Revolution brought with it another, unwanted category of nanoparticles, that is, *industrial* nanoparticles. These are by-products of industrial processes, many of which are quite toxic and damaging to the environment. Common examples include diesel exhaust, welding fume, and industrial combustion emissions. Cigarette smoking, although not strictly speaking an *industrial* process, also contains large numbers of toxic nanoparticles.

The newest category of nanoparticles, and the one for which this book is dedicated, are termed *engineered* nanoparticles (ENPs). Like industrial nanoparticles, these are products of our modern society; the major difference is that this category of nanoparticles is produced specifically to accomplish something, rather than being a by-product. Engineered nanoparticles can range from very simple, such as nanometer-sized alumina used in polymer compounding, to very complex, such as carbon nanotubes (CNTs) and fullerenes.

Exposure Assessment and Safety Considerations for Working with Engineered Nanoparticles,
First Edition. Michael J. Ellenbecker and Candace Su-Jung Tsai.
© 2015 John Wiley & Sons, Inc. Published 2015 by John Wiley & Sons, Inc.

Each category described above will be discussed in some detail in the following sections. First, the basic terms must be defined and discussed.

2.1 NANOTECHNOLOGY, NANOMATERIALS, AND NANOPARTICLES

The three terms that are frequently used interchangeably, but which have distinct definitions, are *nanotechnology*, *nanomaterial*, and *nanoparticle*. Although some confusion and contradictions can be found in the literature, the following are generally accepted consensus definitions of these terms.

2.1.1 Nanotechnology

According to the National Nanotechnology Initiative (NSTC, 2007),

> Nanotechnology is the understanding and control of matter at dimensions approximately between 1 and 100 nm, where unique phenomena enable novel applications. Encompassing nanoscale science, engineering, and technology, nanotechnology involves imaging, measuring, modeling, and manipulating mater at this length scale.

2.1.2 Nanomaterial

Based on the above definition, it follows that a nanomaterial is any material consisting of or containing structures with at least one dimension between 1 and 100 nm.

2.1.3 Nanoparticle

There is now a general agreement in the scientific community that a nanoparticle is an individual particle of matter with at least one dimension in the 1–100 nm size range. The 100 nm diameter given as the upper limit for a particle to be defined as a nanoparticle should not be taken as a strict cutoff. Since aerosols typically have a log-normal size distribution (Hinds, 1999), an aerosol whose median diameter is smaller than 100 nm may well have a significant number of particles larger than 100 nm. In addition, particles in the nanometer size range have very high mobility due to Brownian motion and can agglomerate quickly soon after they are generated.

There is general agreement among scientists that particles in the nanometer size range fall into three categories and are as follows:

1. Naturally occurring nanoparticles These are nanoparticles in the nanometer size range that are found in nature, such as particles released by volcanic eruptions, particles produced by forest fires, salt particles produced by oceanic wave action, and so on.
2. Industrial nanoparticles These are particles in the nanometer size range produced as *unwanted by-products* of our modern industrial society. Common examples include welding fume, diesel exhaust, combustion smoke, and so on.

3. Engineered nanoparticles   These are particles in the nanometer size range produced specifically for a purpose—that is, they are *engineered* to be in the nanometer size range. Engineered nanoparticles range from very simple (e.g., carbon black) to very complex (e.g., drug-delivery nanospheres). They can be spherical (quantum dots, QDs), cylindrical (CNTs), plate-shaped (nanoclay), or irregular (carbon black). In this book, if the term "nanoparticle" or "nanomaterial" is used, it is assumed to mean "engineered nanoparticle" or "engineered nanomaterial" unless stated otherwise.

In addition, the term "nanoparticle powder" (or "nanopowder") is sometimes used. This refers to a bulk material where the individual particle size is in the nanometer range.

2.2   NATURALLY OCCURRING NANOPARTICLES

Luckily for our society, most volcanoes (Fig. 2.1) are dormant at this time. We say "luckily" because the most prominent, and likely most dangerous, naturally occurring nanoparticles are those resulting from volcanic eruptions. This was brought forcefully home with the eruption of the Icelandic volcano Eyjafjallajökull on April 14, 2010, which disrupted air travel over most of Europe for more than a week with the forced cancellation of more than 100,000 flights, due to concerns about jet engine ingestion of volcanic ash. The photo in Figure 2.2, taken by NASA on May 11, 2010, shows the

FIGURE 2.1   The island of Stromboli, a (mostly) dormant volcano. Photo courtesy of Steven Dengler, used with permission, GNU Free Documentation License.

FIGURE 2.2 Icelandic volcano Eyjafjallajökull, May 11, 2010, almost 1 month after eruption. Photo courtesy of the U.S. National Aeronautics and Space Administration.

ash plume still being produced a month after the eruption. Samples of the ash taken on the ground in Iceland immediately after the eruption consisted of metallic ores and were characterized as "…particles range in diameter from approximately 300 μm to a few tens of nanometers … the explosive ash was also remarkable in its abundance of tiny particles attached to larger grains" (Gislason et al., 2011). The German Aerospace Centre (DLR) sent up a test flight over Germany and the Netherlands on April 19, 2010, and found that "…the ash plume is dominated by relatively tiny particles measuring just 0.1 micrometres in diameter, although most of the plume's mass comes from particles measuring 3 micrometres across" (Sanderson, 2010). As we shall see in Chapter 3, this description is very typical of an aerosol dominated by nanometer-sized particles in that the number concentration peaks somewhere in the nanometer size range but most of the mass is contributed by larger particles.

Volcanic eruptions can cause severe acute health effects, including massive death and destruction. Less appreciated are the chronic, long-term environmental and human health impacts of such eruptions. Most people associate active volcanoes with lava flow, but the primary danger from volcanic eruptions is not lava but the large quantity of tephra (solid material ejected from a volcano, including larger rocks and smaller volcanic ash) released to the atmosphere.

Mount Vesuvius is located in Italy about 9 km east of Naples, close to the Bay of Naples. It erupted violently in A.D. 79, and the emitted tephra completely destroyed the Roman cities of Pompeii and Herculaneum, killing an estimated 10,000–25,000 people. The layer of tephra covering Pompeii was estimated to be 3 m deep. More recent volcanic disasters include the destruction of St. Pierre in Martinique in 1902, killing 25,000 people, and Mount St. Helens in the state of Washington, United States, in 1980. These deaths were not, of course, caused by inhaling nanoparticles, but rather were the result of burial in tephra or fires caused by lava flow.

The volcanic ash that contributed to these acute disasters consisted of the larger particles in the eruption; these particles have sufficient mass to settle quickly in the vicinity of the volcano. The smaller, submicrometer-sized ash particles, consisting of minerals, are ejected into the upper atmosphere, where they can remain for months or years. More importantly, volcanic eruptions eject large quantities of sulfur dioxide into the stratosphere; the SO_2 reacts photochemically with water vapor to form submicroscopic sulfate particles that can circulate in the upper atmosphere for months or years. For example, the 1991 Mount Pinatubo eruption is estimated to have ejected 14–20 Tg (15–22 million tons) of SO_2 into the atmosphere. This increased the stratospheric aerosol surface area concentration from 2–5 to 20–100 $\mu m^2/m^3$; a significant fraction of these particles were in the nanometer size range (Pueschel, 1996). Although government agencies have warned the public of the health hazards from inhaling volcanic ash (IVHHN, 2007), actual illnesses are likely to be limited due to the short-term, episodic nature of volcanic eruptions.

The increase in stratospheric aerosols from volcanic eruptions can have a significant effect on the global climate; Pueschel (Pueschel et al., 1994) estimates that the Mount Pinatubo eruption caused a decrease in solar radiation reaching the earth's surface and troposphere of 2.7 W/m². By contrast, the total global warming effect of atmospheric CO_2 buildup since the beginning of the industrial revolution is an increase in solar radiation of about 1.66 W/m² (IPCC, 2007). Going back to the earliest days of the human race, the gigantic eruption of Toba in Sumatra, about 73,000 years ago, created such intense global cooling that experts believe it initiated the last Ice Age and almost wiped out *Homo sapiens*, with perhaps as few as 10,000 people surviving (Rampino and Self, 1992; Zeilinga de Boer and Sanders, 2002).

An important beneficial natural nanoparticle type is the NaCl particles produced by the world's oceans and other salt-containing water bodies. Oceanic wave action produces many small water droplets, which quickly evaporate leaving behind a nanometer-sized NaCl particle. These particles are referred to as condensation nuclei, since when they are carried higher in the atmosphere they serve as condensation sites for water droplets, leading to cloud formation and rain.

2.3 INDUSTRIAL NANOPARTICLES

Occupational and environmental health scientists are very familiar with nanoparticles associated with industrial processes. Nanoparticles have been by-products of such processes since the beginning of the industrial revolution in the late eighteenth and early nineteenth centuries. Industrial nanoparticles are unwanted by-products of beneficial industrial processes. The most common method of industrial nanoparticle production is combustion; diesel exhaust, for example, contains large numbers of particles in the nanometer size range (Fig. 2.3).

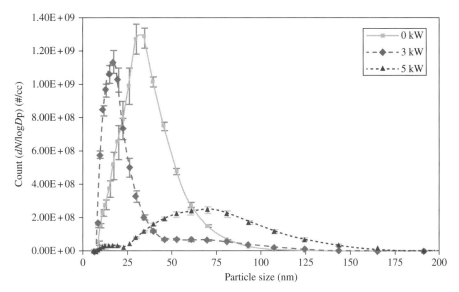

FIGURE 2.3 Particle size distribution in diesel engine exhaust at different engine loads. Reprinted with permission from (Srivastava et al., 2011). Copyright 2011 the Taiwan Association for Aerosol Research.

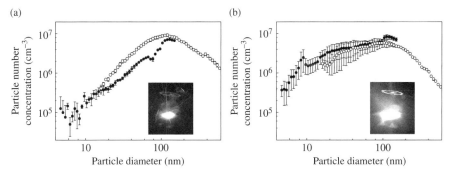

FIGURE 2.4 Scanning Mobility Particle Sizer results for GMAW alloy: (a) during globular transfer mode and (b) spray transfer mode at a sample height 19:2 cm above the arc centerline using both a Nano DMA (darkened circles, 4:53 nm¡dp¡153 nm) and Long DMA (hollow circles, 16:5 nm¡dp¡562 nm) (note: images on each graph are the arcs corresponding to each mode of metal transfer). Reprinted with permission from (Zimmer et al., 2002). Copyright 2002 Elsevier, Inc.

Another common source of industrial nanoparticles is welding. Most of the published literature on welding fume exposures reports the fume size as a mass distribution. Zimmer et al. (Zimmer and Biswas, 2001; Zimmer et al., 2002) used the scanning mobility particle sizer (see Section 7.2 for a description) to measure particle number size distributions for several types of welding operations—a typical result is shown in Figure 2.4. The size distributions appear to be bimodal, with one peak at about 100 nm and a second peak at 10 nm or smaller.

A subject of great concern is welding with high-manganese steel rods, since manganese exposure leads to manganese poisoning with very serious impacts on the central nervous system. Victims of manganese poisoning exhibit symptoms very similar to those associated with Parkinson's disease. Recent research indicates that manganese welding fume, due to its very small size, may pass directly to the central nervous system via the olfactory nerve; this mechanism is described in full detail in Chapter 5.

2.4 ENGINEERED NANOPARTICLES

Engineered nanoparticles are the topic of this book. These are nanoparticles used in manufacturing because of their beneficial properties (as contrasted with industrial nanoparticles, which by definition are unwanted by-products). Engineered nanoparticles can range in structure from very simple to very complex. The simplest nanoparticles are nanoscale particles of materials that are also produced with larger particle sizes. Common examples include carbon black, titanium dioxide, clays, and aluminum oxide (alumina). Figure 2.5 shows a transmission electron micrograph (TEM) photo of nanoalumina. This particular nanoalumina has a typical polydisperse size distribution, with a count median diameter of about 40 nm. The TEM image shown is of a sample of the bulk material; as shown, the individual particles typically form irregular agglomerates with a largest dimension of 400 nm or so.

The use of these very simple engineered nanomaterials is typically straightforward; for example, nanoalumina is used as a filler in the manufacture of polymer

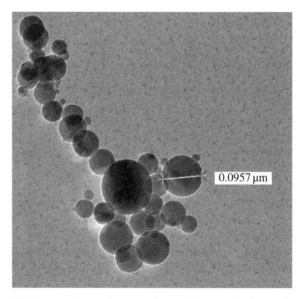

FIGURE 2.5 TEM image of an agglomerate from a bulk sample of nanoalumina. Measured particle is approximately 100 nm in diameter. Image by S. Tsai.

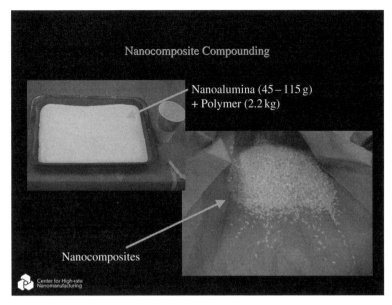

FIGURE 2.6   Photos from the UMass Lowell Plastics Engineering laboratory, showing the bulk nanoalumina used as a filler and the resulting nanocomposite. Photos by S. Tsai.

nanocomposites (Fig. 2.6). The larger surface-to-volume ratio of alumina in the nanometer size range, as compared to larger particles, improves the mechanical properties that the nanoalumina filler imparts to the polymer.

Much current interest centers on the more exotic and complex engineered nanoparticles, such as CNTs, fullerenes or C_{60}, and QDs. A great deal of research is now being performed worldwide—both on the manufacturing of the particles themselves and their incorporation into useful devices. Much excitement has been generated in recent years on the potential for these exotic nanoparticles to revolutionize manufacturing, with applications as diverse as mechanical structures, electronics, energy storage, and drug delivery.

2.4.1   Carbon Nanotubes

CNTs (Fig. 2.7) were first reported by Soviet scientists in the 1950s (Monthioux and Kuznetsov, 2006), but their discovery is generally attributed to Sumio IIjima of NEC Corp. in 1991 (Iijima, 1991), and are allotropes of carbon that are formed into a cylindrical structure. The sheet that makes up the cylindrical surface is one atom thick and when it is in a flat form is called graphene. During the manufacturing process, CNT can form a single-walled (SWCNT) (Fig. 2.7a and b) or a multiwalled structure (MWCNT) (Fig. 2.7c). In addition, the tubes can "roll up" at different angles relative to the graphene lattice structure, and this angle affects the tube's electrical and mechanical properties. A SWCNT usually has a diameter of about 1 nm, and the length is highly variable but can be in the millimeter range. A MWCNT

(a) (b)

(c)

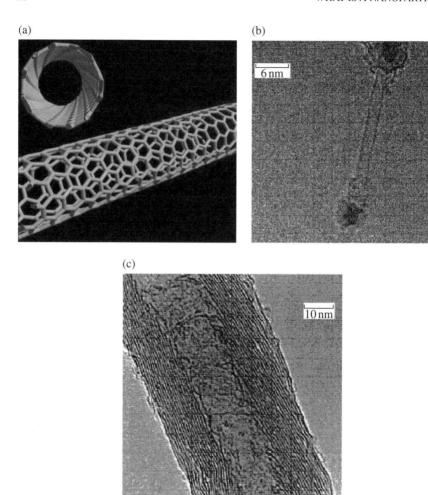

FIGURE 2.7 Carbon nanotubes: (a) idealized single-walled carbon nanotube (Image courtesy of the National Cancer Institute); (b) photomicrograph of a single-wall carbon nanotube (Image courtesy of the U.S. National Aeronautic and Space Administration); and (c) photomicrograph of a multiwalled carbon nanotube (Image courtesy of the U.S. National Aeronautic and Space Administration).

consists of concentric cylinders of graphene and consequently has a larger diameter than SWCNTs. There are several CNT manufacturing processes including chemical vapor deposition (CVD), laser ablation, and arc discharge; CVD requires the use of a metal catalyst to initiate the cylinder formation, complicating the toxicity evaluation of the resulting nanoparticles.

Closely related to CNTs are carbon nanofibers (CNFs), which are also hollow cylinders of carbon atoms. The difference between the two is that CNTs consist of cylindrical sheets whereas CNFs have a range of more heterogeneous structures,

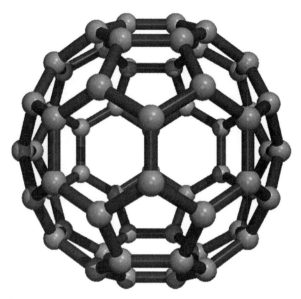

FIGURE 2.8 C_{60} fullerene. Image courtesy of U.S. Oak Ridge National Laboratory.

such as "stacked cups." CNFs have reduced mechanical properties compared to CNTs, but they can be made much more cheaply and thus have many potential applications where the superior properties of CNTs are not needed.

2.4.2 Fullerenes

Fullerenes are usually thought of as hollow spheres composed entirely of carbon atoms (Fig. 2.8), although CNTS are also categorized as fullerenes. The most common form of spherical fullerene consists of 60 carbon atoms and is also called C_{60} or buckminsterfullerene (named after Richard Buckminster Fuller, famous designer of geodesic domes which resemble C_{60}). C_{60}, the smallest fullerene, also resembles a soccer ball, with 20 hexagons and twelve pentagons in an interlocking structure. The diameter of C_{60} is about 1 nm; strictly speaking, a fullerene (and a CNT, for that matter) is not a nanoparticle but rather a single molecule consisting of 60 carbon atoms. A fullerene is fairly inert, so chemists frequently add chemical functional groups at one or more of the carbon atoms, in order to allow individual fullerenes to be combined into larger structures or to be attracted to receptors on cells. This also complicates toxicology studies, since each functionalized fullerene is a new material with its own properties.

2.4.3 Quantum Dots

QDs, which are a category of NPs made from semiconductor crystals, have gained much attention over the past few decades (Dabbousi et al., 1997; Danek et al., 1996; Medinitz et al., 2005). They are very small in diameter ranging from

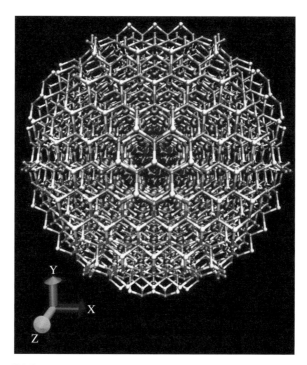

FIGURE 2.9   Cd–Se quantum dot structure. The cadmium and selenium atoms make up the lattice structure, surrounding a cloud of electrons in their excited state. Image by Sebastien Hamel/U.S. Lawrence Livermore National Laboratory.

1 to 10 nm (Bae et al., 2009; Empedocles et al., 1996; Guo et al., 2007; Lovric et al., 2005; Medinitz et al., 2005). They are also referred to as the artificial atom, since the electrons present in QDs occupy discrete energy levels similar to atoms. QDs typically are smaller than the excitation Bohr radius, which results in a phenomenon known as quantum confinement—hence the name. These nanocrystals are made of atoms that belong to Group II–VI or III–V elements in the periodic table, such as cadmium and selenium (Cd–Se) or indium and phosphorous (In–P) (Howarth et al., 2005; Medinitz et al., 2005). An example of a Cd–Se QD is shown in Figure 2.9.

An important property of QDs is that they absorb photons over a range of frequencies and reradiate photons at a discrete frequency that is a function of their diameter. Compared to fluorescent dyes, QDs exhibit bright fluorescence, high photostability, greater quantum yields, and wide absorption spectra, which allow colors to be excited at a single excitation wavelength (Alivisatos, 1996). Although many fluorescent dyes exhibit these same characteristics as QDs, they undergo photo bleaching, unlike QDs (Howarth et al., 2005; Law et al., 2009). These properties make QDs ideal for use in biomedical research, which includes *in vivo* cell imaging, targeted drug delivery, and diagnostics (Li et al., 2009).

2.5 EMERGING USES FOR ENGINEERED NANOPARTICLES

The range of potential uses for ENPs is truly staggering, encompassing the mundane (consumer products that incorporate ENPs purely as a marketing gimmick) to potentially life-changing technologies such as nanomedicine. Although the purpose of this book is to discuss some of the potential adverse aspects of nanotechnology, it is important to recognize the very real potential benefits to society that are likely to result from the use of these materials.

The Woodrow Wilson International Center for Scholars Project on Emerging Nanotechnologies maintains a database of web sites that advertise commercially available nanoparticle-containing products (WWICS, 2012); as of March 2011 (the last time the inventory was updated), there were more than 1300 products included in the inventory. The majority of products in the database (738) were in the category "Health and Fitness," which includes the subcategories personal care products (267), clothing (182), cosmetics (143), sporting goods (119), filtration devices (43), and sunscreens (33). The next most popular categories were "Home and Garden" with 209 entries and "Automotive" with 126. A quick review of the database reveals that most of the advertised products are decidedly "low tech," as contrasted with the "high tech" applications that are the focus of ongoing research. In fact, it is clear that many of the currently available products that advertise themselves as having some connection to nanotechnology are doing so only as a marketing gimmick.

The worldwide research into nanotechnology is moving beyond the use of ENPs as a marketing gimmick and toward their truly revolutionary use in a wide variety of applications. Among these are medical applications (drug delivery, imaging, etc.), electronics, energy, and advanced composite materials. Two examples of such applications will be discussed briefly to give the reader an idea of the range and likely impacts of these exciting new uses of ENPs.

One of the first commercial applications of QDs is for lighting. Light-emitting diodes (LEDs) are much more efficient than incandescent or even compact fluorescent lights (CFLs) and have lifetimes of many years, making them a highly desirable method for commercial and residential lighting. Their widespread adoption would have a significant positive impact on energy use and atmospheric CO_2 emissions. Unfortunately, however, the wavelength pattern from an LED is such that the light appears "harsh" compared to incandescent bulbs. A Massachusetts company[1] has marketed a product where an LED array is coated with a layer of QDs; the QDs absorb the LED photons and reradiate them in a "softer" pattern that matches incandescent lights, with no loss in energy efficiency.

In the area of mechanical structures, a New Hampshire company[2] has developed the technology to produce very long, pure CNTs that are made into yarns or nonwoven sheets as they leave the CNT furnace. These products have many advantageous properties over traditional alternative materials, such as Kevlar®—they have high

---

[1] QD Vision, Inc., Lexington, MA.
[2] Nanocomp Technologies, Inc., Merrimack, NH.

electrical conductivity and high mechanical strength, meaning they can be used in applications as diverse as bulletproof vests and electromagnetic interference (EMI) shielding for missiles and planes—at a much lower weight than the current products. Importantly from an environmental health and safety perspective, this production process minimizes the probability that free CNTs will escape their enclosed and ventilated furnaces.

The tension between the potential benefits of nanotechnology and the possible adverse health and environmental effects of the various ENPs is the subject of much intellectual interest at this time. For example, Shatkin (2008) writes:

> The commercialization of nanotechnology is literally under the microscope. Numerous non-governmental organizations that focus on environmental health and consumer issues currently are calling for a moratorium on all products containing nanotechnology until their safety and risks are known…Regulatory agencies are being asked to develop standards, yet the data are not currently available to ascertain safe levels of many new materials….

> Demands to understand and address risks in real time, that is, during development, add a difficult dimension to nanotechnology development … the understanding of behavior at the nanoscale is in very early stages, and it is premature to make long-lasting decisions about nanotechnology without this understanding. However, now is the time to begin the analysis, while the actual risks from nanomaterials are small, because they are produced in low levels and very few people are exposed in very small amounts, to guide decision making for when they are in widespread use. Addressing potential exposures now is the best way to mitigate the long-term risks of nanomaterials and nanotechnology that are currently unknown.

2.6 OTHER USEFUL DEFINITIONS

Several other terms are used throughout the book; their definitions are as follows.

2.6.1 Aerosol

An aerosol is defined as a suspension of particles in a gas (Hinds, 1999). The particles can be solid and liquid, and in our case the gas will be air. The term "suspension" implies that the particles will remain suspended in air for some period of time. Once a particle is suspended in air, it is subject to two forces, that is, gravity and aerodynamic drag; the force of gravity is constant, but the drag force is proportional to the particle's velocity. Once the particle accelerates to a velocity where the gravity and drag forces are equal, the particle settles at a constant velocity, called the terminal settling velocity. For particles in the micrometer size range and smaller, so-called Stokes drag predominates and the terminal settling velocity is

$$V_{TS} = \frac{\rho_p d_p^2 g C_c}{18\eta} \tag{2.1}$$

where V_{TS} = terminal settling velocity (m/s), ρ_p = particle density (kg/m^3), d_p = particle diameter (m), g = acceleration due to gravity = 9.8 m/s^2, C_c = Cunningham correction factor[3] (dimensionless), η = air viscosity = 1.81×10^{-5} Pa·s at STP.

Table 2.1 lists the terminal settling velocities of water droplets of different diameters, along with the time each particle would take to settle 2 m. Water typically is used in illustrations like this because it has a "reference" density of 1000 kg/m^3; most solid materials have densities two to three times that of water, so their settling velocities will be proportionally higher. As a general rule, the upper size limit for particle diameters typically considered to meet the definition of "aerosol" is about 100 μm since they settle out of the air fairly quickly, as shown in Table 2.1. As also shown in the table, particles meeting the definition of a nanoparticle ($d \leq 100$ nm) have such low settling velocities that they remain airborne for very long periods; the time it takes for a 1 nm particle to settle 2 m, given in the table as 5×10^6 min, is approximately 10 years! Nanoparticles suspended in air remain there indefinitely and essentially act very much like the gas molecules surrounding them; their primary motion is due to the bulk motion of that air.

2.6.2   Particle Inertia

Inertia is the tendency of an object to maintain a constant velocity when subject to nonzero net forces. The more massive an object, the more inertia it has. This holds for aerosol particles as for all other objects; larger aerosol particles in motion tend to move in straight lines when subject to external forces and are said to have a higher inertia than smaller particles. The inertial behavior of a particle is also referred to as its *aerodynamic* behavior, since we are talking about motion through a gas.

The concept of aerodynamic diameter is used to characterize the inertial behavior of particles of unknown shape and density. If two particles have the same aerodynamic diameter, that means they behave exactly the same when subject to inertial forces. By convention, all particle behavior is normalized to water droplets, which serve as an inertial reference; thus, the aerodynamic diameter of a particle is the diameter of the

TABLE 2.1   Settling velocities and time to settle 2 m for water droplets of various diameters

Droplet diameter (μm)	Settling velocity (m/s)	Time to settle 2 m (min)
0.001	7×10^{-9}	5×10^6
0.01	7×10^{-8}	5×10^5
0.1	9×10^{-7}	4×10^4
1	4×10^{-5}	8×10^2
10	3×10^{-3}	10
100	3×10^{-1}	0.1

Adapted from Hinds (1999).

[3] The Cunningham correction factor accounts for departures from Stokes law for very small particles, where the air can no longer be considered a continuous medium. It is equal to 1.0 for large particles and increases to 224 for a 1 nm particle.

water droplet with the same inertial behavior as the particle. As indicated by their settling velocities, which are essentially zero, NPs do not move inertially with respect to the gas surrounding them; thus, the concept of aerodynamic diameter for a nanoparticle has little validity. In fact, it is next to impossible to measure the aerodynamic diameter of a NP. The actual physical diameter and shape of a NP are much more important descriptors of their behavior than aerodynamic diameter.

2.6.3 Brownian Motion

Since nanoparticles must deposit on lung surfaces or the skin in order to cause an adverse health effect and airborne nanoparticles do not settle significant distances due to gravity, the reader may wonder why, since airborne nanoparticles do not settle significant distances due to gravity, we have any concern about exposure to them. The answer is that there is another force acting on nanometer-sized particles that causes them to move relative to the air, and that motion is called Brownian motion.

The force causing Brownian motion is due to the interaction of aerosol particles with the surrounding air molecules. Air molecules move rapidly in random directions due to their thermal energy, and they make elastic collisions with each other and with any object they encounter. For large airborne objects, such as baseballs and micrometer-sized particles, at any instant the average number of collisions on all sides is essentially the same, and the force imparted by any one collision is very small compared to the mass of the object. As aerosol particles get smaller, there is a greater likelihood that any instant there will be more collisions on one side of the particle than the other, and the force of each collision is larger relative to the particle mass. When this happens, the particle gets a "push" in a random direction, followed an instant later by a "push" in another random direction. This results in the random movement of the particle through the air, a phenomenon first noted in 1827 by botanist Robert Brown for single-celled organisms in water (Hinds, 1999).

Just as with molecules in water or air, the random movement of aerosol particles due to Brownian motion causes a net transfer of particles from a region of high concentration to a region of low concentration; this motion is characterized by the particle's *diffusion coefficient*. Diffusion was first studied and understood for mixtures of two gases; early in the twentieth century, it was known that the same principle applied to aerosol particles suspended in air, but the equation for predicting a particle's diffusion coefficient was not known. It is interesting (at least to aerosol scientists!) that Albert Einstein, at the same time he was doing his revolutionary work on special relativity, was engaged in theoretical aerosol physics. While it is true that he was awarded the Nobel Prize in physics for his work on relativity, one of his three 1905 papers cited by the Nobel committee was his development of the theoretical equation for the particle diffusion coefficient, the so-called Stokes–Einstein equation:

$$D = \frac{kTC_c}{3\pi\eta d_p} \tag{2.2}$$

where D = particle diffusion coefficient (m²/s), k = Boltzmann's constant $(1.38 \times 10^{-23} \text{J/K})$, T = gas temperature (K), η = viscosity (Pa·s), and C_c = Cunningham correction factor, dimensionless.

As indicated by Equation (2.2), at standard conditions (constant temperature and viscosity), a particle's diffusion coefficient depends only on its diameter and, as expected, the diffusion coefficient increases as the diameter decreases. Since Brownian motion is random, a collection of aerosol particles of the same diameter will be displaced at different distances along an arbitrary axis in any given time interval. The root mean square (rms) average distance a collection of particles moves in time t is given by:

$$x_{rms} = \sqrt{2Dt} \qquad (2.3)$$

where x_{rms} = root mean square displacement (m).

Hinds (1999) presents an interesting comparison of the settling distance in 1 s versus the rms displacement in 1 s for various diameter particles, repeated here in Table 2.2. As seen in the table, a 1 nm particle travels almost 50,000 times further, on average, due to Brownian motion than gravity settling, whereas a 100 µm particle travels 500,000 times further due to gravity settling compared to its Brownian motion. This leads to the conclusion that nanoparticle motion is dominated by Brownian diffusion, while the motion of particles larger than 1 µm is dominated by inertial effects. It is also interesting to note that particles with diameters between 100 nm and 1 µm, the so-called "submicrometer" particles, have very little motion from *either* effect; this helps explain their minimal deposition in the human respiratory system (see Section 4.2) and in filters (see Section 10.1).

2.6.4 Particle Diameter

Aerosol particles, including nanoparticles, are most typically characterized by referring to their diameter. Of course, "diameter" implies that the particle is a sphere, but in actuality many aerosol particles have a shape that is far from spherical. Liquid

TABLE 2.2 Net displacement of water droplets in 1 s due to Brownian motion and gravity settling

Droplet diameter (µm)	Root mean square (rms) Brownian displacement (m)	Gravity settling distance (m)	Ratio
0.001	3.3×10^{-3}	7×10^{-9}	48,000
0.01	3.3×10^{-4}	7×10^{-8}	4,800
0.1	3.7×10^{-5}	9×10^{-7}	40
1	7.4×10^{-6}	4×10^{-5}	0.2
10	2.2×10^{-6}	3×10^{-3}	7×10^{-4}
100	6.9×10^{-7}	3×10^{-1}	2×10^{-6}

Adapted from Hinds (1999).

aerosol particles and specially manufactured solid test aerosol particles such as polystyrene latex spheres are spherical, as are some ENPs (e.g., quantum dots). Solid aerosol particles have a wide variety of shapes, ranging from geometric structures such as cubes (e.g., NaCl particles) and truncated icosahedrons (C_{60} fullerenes), irregular (particles formed by mechanical action such as cutting, grinding, etc.), flat plates (nanoclay), branchlike collections of molecules (fume particles formed by condensation from the vapor phase), and fibrous (carbon nanotubes).

Given this range of shapes, what can we make of the term "diameter"? To some extent, it depends on the context in which it is used. For example, when discussing measurements made by automated particle counting and sizing instruments, diameter refers to the size the instruments *senses* the particle to be; the relationship between the measured diameter and the particle's actual diameter depends on the measurement method used by that particular instrument. This issue is discussed in detail in Chapter 7. When sizing irregularly shaped particles under the microscope (either light or electron), various conventions have been developed, the most common of which is the projected area diameter, defined as the diameter of the circle that has the same projected area as the particle being sized (Hinds, 1999).

In the most general sense, the term diameter is used more casually than the examples above and can be thought of as the approximate size of an irregular particle. This convention works satisfactorily for most particles, with the exception of particles such as fibers, where it is preferable to use length and diameter of the cylinder approximating the fiber to describe the particle size.

2.6.5 Agglomerate versus Aggregate

Although ENPs must exist as individual particles when they are formed, they can quickly form collections of particles. It is always the case, for example, that the individual structures of a bulk nanopowder will be collections of particles; such collections are called agglomerates or aggregates. As an example, in our laboratory we frequently use a particular nanoalumina as a test material. The individual particles have a nominal diameter of about 30–50 nm, but the bulk material, when examined under electron microscopy, consists of agglomerates about 200 nm in diameter.

There is considerable confusion in the literature as to which term, agglomerate or aggregate, is appropriate for describing such collective particle structures. Most sources distinguish between them based on how tightly the particles are held together; unfortunately, different groups define them in completely contradictory ways. It appears that most government agencies are consistent in their definitions. For example, the European Union defines an agglomerate as "a collection of weakly bound particles or aggregates where the resulting external surface area is similar to the sum of the surface areas of the individual components" and an aggregate as "particle comprising of strongly bound or fused particles," (Maynard, 2011); this is consistent with the definitions of the US National Institute of Standards and Technology (Jillavenkatesa et al., 2001).

Peer-reviewed articles, however, show considerable confusion. Among many other examples, Balazy and Podgorsky (2007) discuss "fractal-like aggregates emitted by diesel engines" while Maynard and Keumpel (2005) discuss "changes in particle diameter through agglomeration." Kuhlbusch and Fissan (2006) describe the manufacturing process for carbon black as follows:

> In the first process phase, chemical reactions take place in the reactor to produce the primary carbon black particles during partial combustion or thermal decomposition of hydrocarbons by gas-to-particle conversion. Primary particle sizes can range from 1–500 nm, with most produced in the 10 nm to 100 nm range. A few to many tens of primary particles immediately form highly branched chains of carbon black called aggregates. During the process of collection, these aggregates are agglomerated in cyclones and bag houses to much larger entities.

In this case, the aggregates formed in the furnace are more-or-less permanent structures, while the agglomerates can be broken apart by the application of energy. Our unscientific review of the literature indicates that a majority of scientists and organizations use this distinction, that is, aggregates are more-or-less permanent collections of primary particles formed when the particles were created, while agglomerates a more loosely-held collections of primary particles and/or aggregates formed later in the particles' life, such as during bulk storage.

Nichols et al. (2002) attempted to summarize the confusion; after first stating "the terms agglomerate and aggregate are qualitative and have been interchanged by most researchers (without any thought as to what they actually mean) for so long that it probably no longer matters how they are used," and after further discussion they conclude "we therefore propose that for powders, use of the term 'aggregate' is discontinued and the term 'agglomerate' is used exclusively." We will follow this convention and use the term aggregate only to indicate the type of particle referenced in the carbon black industry, that is, a collection of particles formed during the manufacturing process that are difficult to break apart by the application of energy; this convention is consistent with the government agency definitions cited above.

2.7 SUMMARY

Nanoparticles are not new, but engineered nanoparticles are. They have started to have revolutionary impacts on many aspects of our society. While the beneficial impacts are to be applauded and encouraged, we must pay close attention to the possible negative impacts as well. As we hope will become clear through the remainder of this book, addressing those negative impacts has some aspects that are straightforward and some that are complex. It is straightforward because ENPs are small particles, and the health and safety community has been evaluating and controlling exposures to small particles for more than a century. It is complex because ENPs are *very* small particles. Moving from the micrometer scale to the nanometer scale has *scientific* implications for all aspects of health and safety, from toxicology to exposure characterization to exposure control, as well as *policy* implications such as the establishment worker exposure and industrial emissions standards. These are the issues addressed in the remainder of this book.

REFERENCES

Alivisatos AP. Perspectives on the physical chemistry of semiconductor nanocrystals. J Phys Chem 1996;100:13226–13239.

Bae PK, Kim KN, Lee SJ, Chang HJ, Lee CK, Park JK. The modification of quantum dots probes used for the targeted imaging of his-tagged fusion proteins. Biomaterials 2009;30:836–842.

Balazy A, Podgorski A. Deposition efficiency of fractal-like aggregates in fibrous filters calculated using Brownian dynamic method. J Coll Interf Sci 2007;311:323–327.

Dabbousi BO, Rodriguez VJ, Mikulec FV, Heine JR, Mattoussi H, Ober R, Jensen KF, Bawendi MG. (CdSe)ZnS core-shell quantum dots: synthesis and characterization of a size series of highly luminescent nanocrystallites. J Phys Chem B 1997;101:9463–9475.

Danek M, Jensen KF, Murray CB, Bawendi MG. Synthesis of luminescent thin-film CdSe/ZnSe quantum dot composites using CdSe quantum dots passivated with an overlayer of ZnSe. Chem Mater 1996;8:173–180.

Empedocles SA, Norris DJ, Bawendi MG. Photoluminescence spectroscopy of single CdSe nanocyrstallite quantum dots. Phys Rev Lett 1996;77:3873–3876.

Gislason SR, Hassenkamb T, Nedel S, Bovet N, Eiriksdottir ES, Alfredsson HA, Hem CP, Balogh ZI, Dideriksen K, Oskarsson K, Sigfusson B, Larsen G, Stipp SLS. Characterization of Eyjafjallajökull volcanic ash particles and a protocol for rapid risk assessment. PNAS 2011;108:7307–7312.

Guo G, Liu W, Liang J, He Z, Xu H, Yang X. Probing the cytotoxicity of CdSe quantum dots with surface modification. Mater Lett 2007;61:1641–1644.

Hinds WC. *Aerosol technology—properties, behavior, and measurement of airborne particles*. New York: Wiley-Interscience; 1999.

Howarth M, Takao K, Hayashi Y, Ting AY. Targeting quantum dots to surface proteins in living cells with biotin ligase. PNAS 2005;102:7583–7588.

Iijima S. Helical microtubules of graphitic carbon. Nature 1991;354:56–58.

Intergovernmental Panel on Climate Change [IPCC]. *Fourth assessment report: Climate change 2007* Geneva: Intergovernmental Panel on Climate Change; 2007.

International Volcanic Health Hazard Network [IVHHN]. *The Health Hazards of Volcanic Ash—A Guide for the Public*. Durham: IVHHN; 2007.

Jillavenkatesa A, Dapkunas SJ, Lum LSH. *Particle size characterization Special Publication 960-1*. Gaithersburg (MD): U.S. Department of Commerce, National Institute for Standards and Technology; 2001.

Kuhlbusch TAJ, Fissan H. Particle characteristics in the reactor and pelletizing areas of carbon black production. J Occup Environ Hyg 2006;3:558–567.

Law WC, Yong KT, Roy I, Ding H, Hu R, Zhao W, Prasad PN. Aqueous-phase synthesis of highly luminescent CdTe/ZnTe core/shell quantum dots optimized for targeted bioimaging. Small 2009;5:1302–1310.

Li R, Dai H, Wheeler TM, Sayeeduddun M, Scardino PT, Frolov A, Ayala GE. Prognostic value of akt-1 in human prostate cancer: a computerized quantitative assessment with quantum dot technology. Clin Cancer Res 2009;15:3568–3573.

Lovric J, Cho SJ, Winnik FM, Maysinger D. Unmodified cadmium telluride quantum dots induce reactive oxygen specs formation leading to multiple organelle damage and cell death. Chem Biol 2005;12:1227–1234.

Maynard AD. 2011. EC adopts cross-cutting definition of nanomaterials to be used for all regulatory purposes. In: 2020 Science. Available at http://2020science.org/2011/10/18/ec-adopts-cross-cutting-defintion-of-nanomaterials-to-be-used-for-all-regulatory-purposes/. Accessed October 18, 2011.

Maynard AD, Kuempel ED. Airborne nanostructured particles and occupational health. J Nanopart Res 2005;7:587–614.

Medinitz IL, Uyeda HT, Goldman ER, Mattoussi H. Quantum dot bioconjugates for imaging, labelling and sensing. Nat Mater 2005;4:435–446.

Monthioux M, Kuznetsov VL. Who should be given the credit for the discovery of carbon nanotubes? Carbon 2006;44:1621–1623.

National Science and Technology Council [NSTC]. *The National Nanotechnology Initiative: Strategic Plan.* Washington (DC): NSTC, the National Nanotechnology Initiative, Executive Office of the President of the United States; 2007.

Nichols G, Byard S, Bloxham MJ, Botterill J, Dawson NJ, Dennis A, Diart V, North NC, Sherwood JD. A review of the terms agglomerate and aggregate with a recommendation for nomenclature used in powder and particle characterization. J Pharm Sci 2002;91:2103–2109.

Pueschel RF. Stratospheric aerosols: formation, properties, effects. J Aerosol Sci 1996;27:383–402.

Pueschel RF, Russell PB, Allen DA, Ferry GV, Snetsinger KG, Livinston JM, Verma S. Physical and optical properties of the Pinatubo volcanic aerosol: aircraft observations with impactors and a sun-tracking photometer. J Geophys Res 1994;99:12915–12922.

Rampino MR, Self S. Volcanic winter and accelerated glaciation following the Toba super-eruption. Nature 1992;359:50–52.

Sanderson K. Questions fly over ash-cloud models. Nature 2010;464:1253.

Shatkin JA, editor. Introduction: assessing nanotechnology health and environmental risks. In: *Nanotechnology—Health and Environmental Risks.* Boca Raton (FL): CRC Press; 2008. p 12–13.

Srivastava DK, Agarwal AK, Gupta T. Effect of engine load on size and number distribution of particulate matter emitted from a direct injection compression ignition engine. Aerosol Air Qual Res 2011;11:915–920.

Woodrow Wilson International Center for Scholars [WWICS]. 2012. The project on emerging technologies. WWICS. Available at www.nanotechproject.org/. Accessed November 22, 2012.

Zeilinga de Boer J, Sanders DT. *Volcanoes in Human History: The Far-Reaching Effects of Major Eruptions.* Princeton (NJ): Princeton University Press; 2002.

Zimmer AT, Biswas P. Characterization of the aerosols resulting from arc welding processes. J Aerosol Sci 2001;32:993–1008.

Zimmer AT, Baron PA, Biswas P. The influence of operating parameters on number-weighted aerosol size distribution generated from a gas metal arc welding process. J Aerosol Sci 2002;33:519–531.

3

WHY ARE WE CONCERNED? THE UNIQUE PROPERTIES OF NANOPARTICLES

As discussed in the Chapter 2, there are many properties of engineered nanoparticles (ENPs) that lead to excitement about their potential uses to the benefit of society. At the same time, however, it is becoming increasingly clear that there are also environmental and human health concerns regarding the use of these particles. The toxicology evidence for this is reviewed in Chapter 5; here, we will discuss the factors that are inherent to nanoparticles that might lead to this concern.

There is strong evidence that the toxicity of many particle types increases when the particle size shrinks from the micrometer range to the nanometer range. For example, a recently published study (Jiang et al., 2009) compared the bacterial toxicity of nanometer-sized aluminum, silicon, titanium, and zinc oxides with their micrometer-sized counterparts. All of the nanometer-sized particles, with the exception of titanium oxide, showed greater toxicity. What might cause this difference? The likely candidates are discussed below.

3.1 SURFACE-TO-VOLUME RATIO

It is clear that the basic *chemical* properties of the material do not change as particle size changes but the basic *physical* properties do change. The physical property most likely associated with increased toxicity is the increasingly important role of *surface area* as particles become smaller.

Exposure Assessment and Safety Considerations for Working with Engineered Nanoparticles,
First Edition. Michael J. Ellenbecker and Candace Su-Jung Tsai.
© 2015 John Wiley & Sons, Inc. Published 2015 by John Wiley & Sons, Inc.

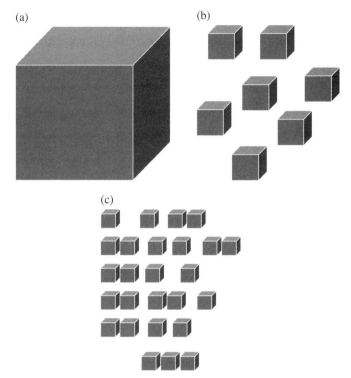

FIGURE 3.1 Graphite cubes: (a) 1 m on a side; (b) 1 μm on a side; and (c) 1 nm on a side (not to scale).

We can demonstrate this with a simple thought experiment, illustrated in Figure 3.1. Let's assume we have a solid block of graphite (a form of carbon) in the form of a single cube, 1 m on a side (Fig. 3.1a). Clearly, this is not an aerosol, but we are going to make it one in our next step. We can do some basic calculations about its fundamental characteristics. First, its *volume* is, of course, 1 m³; since the density of graphite is approximately 2200 kg/m³, its *mass* is 2200 kg. The *surface area* of the block is 6×1 m², or 6 m². All these numbers are entered in the first line of Table 3.1, along with the calculated surface area to volume ratio of 6 m⁻¹.

Now we will continue the thought experiment by using a theoretical laser saw that can very precisely cut the large block (without creating any waste); we will cut it 1000 times along each dimension, creating smaller cubes that now are each 1 mm on a side. Since we cut along three axes, we created $10^3 \times 10^3 \times 10^3 = 10^9$ 1 mm cubes. Since the total volume of all the cubes is still 1 m³, the volume of each cube must be 10^{-9} m³; this is entered in the second line of Table 3.1, along with calculations of the other properties. Notice that the total surface area of all the particles, and the surface-to-volume ratio of each particle, has increased by a factor of 1000.

If we now cut each of the 1 mm cubes 1000 times along each dimension, we end up with 10^{18} cubes with a dimension of 1 μm on each side (Fig. 3.1b). Once again, the

TABLE 3.1   Properties of graphite cubes of various dimensions

		Amount per particle				
Cube length (m)	Number of cubes	Volume (m^3)	Mass (kg)	Surface area (m^2)	Surface/ volume ratio (m^{-1})	Total surface area (m^2)
1	1	1	2200	6	6	6
10^{-3}	10^9	10^{-9}	2.2×10^{-6}	6×10^{-6}	6×10^3	6×10^3
10^{-6}	10^{18}	10^{-18}	2.2×10^{-15}	6×10^{-12}	6×10^6	6×10^6
10^{-9}	10^{27}	10^{-27}	2.2×10^{-24}	6×10^{-18}	6×10^9	6×10^9

total surface area and particle surface-to-volume ratio have increased by a factor of 1000. Finally, we can make some nanoparticles. If we cut each micrometer-sized cube 1000 times along each dimension, we now will have 10^{27} 1-nm nanocubes (Fig. 3.1c). The first thing to contemplate is the very large number of nanoparticles we have produced. One octillion (10^{27}) is a *very* large number. For example, 10^{27} US nickels have approximately the same total mass as the earth! More dramatically perhaps, if we continue our thought experiment and pile all the nanocubes in a single stack, each one is 10^{-9} m high and there are 10^{27} of them, so the stack's height will be 10^{18} m. This is another number that is hard to grasp, but it turns out to be about equal to 100 light years, and 25 times greater than the distance to Alpha Centauri, our nearest neighboring star!

The next thing to note is the very small mass of each nanoparticle. The mass of one of our nanocubes equals the mass of a water droplet with about 100 water molecules. This illustrates one basis for the frequently made statement, discussed in Chapter 2, that nanoparticles are so small that they approach the molecular level and behave very much like molecules when suspended in air or water.

If we next consider surface area, the total surface area of all of our nanocubes (6×10^9 m^3 or 6000 km^2) is one billion times that of our starting cube and is about equal to the size of the US state Delaware. More importantly from a toxicity viewpoint, the surface-to-volume ratio has increased from 6 m^{-1} for our original graphite block to 6×10^9 m^{-1} for each of our nanoparticles. At the same time, the mass of each nanoparticle has become almost vanishingly small (2.2×10^{-21} g). This is the basis for scientists' concern that toxicological endpoints that depend on surface area and/or number are likely to be much more important for nanoparticles than endpoints that are a function of particle mass.

3.2   PARTICLE SIZE

The particle size itself likely contributes to differences in toxicity. It is obvious, but needs emphasizing, that a 10 nm diameter particle is 1000 times smaller than a 10 μm diameter particle. Figure 3.2 compares a scanning microscope image of asbestos fibers (a) to a transmission electron microscopy image of carbon nanofibers (b). Note

(a)

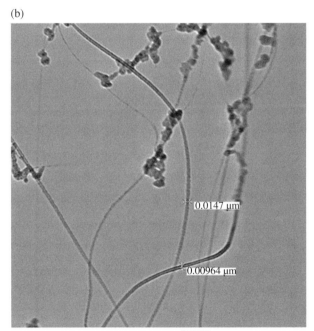

10 μm (0.0004″)

(b)

0.0147 μm

0.00964 μm

nanocomp source - 4 – 1 0608′.tif   100 nm
CNT source 0608′ - 4 – 1   HV = 100kV
Cal: 955.975 pix/micron   Direct Mag: 20000x

TEM Mode: Imaging
Microscopist: Candace

FIGURE 3.2 (a) Asbestos fibers as seen under scanning electron microscopy. Image courtesy of US Environmental Protection Agency and (b) carbon nanofibers as seen under transmission electron microscopy. Image taken by S. Tsai.

the scale lines on each figure, comparing the $10\,\mu m$ (10,000 nm) line in (a) to the 100 nm line in (b). The fibers look very similar, but the carbon nanotubes (CNTs) are about 100 times smaller in diameter than the asbestos fibers! A particle's size can contribute in several ways to its toxicity, including the deposition pattern of the particle in the respiratory system, the likelihood of the particle to penetrate the skin, and particle mobility once it enters the body. Referring to our example in Figure 3.2, the deadly disease mesothelioma is a cancer of the pleural lining of the lung and is associated with asbestos exposure. This occurs when asbestos fibers first deposit in the alveolar region of the lung and then migrate through the cell walls into the pleural cavity. CNTs, being orders of magnitude thinner than asbestos fibers, may more easily make this migration; this is discussed in Section 5.4.

3.3   PARTICLE CONCENTRATION

One might wonder how the admittedly unrealistic thought experiment described above relates to actual releases of nanoparticles during a typical industrial process. The important point to keep in mind is that very small releases of nanometer-sized material on a mass basis will correspond to very large numbers of particles. Again, let's look at an example. In industrial settings, airborne concentrations when measured on a mass basis are typically in the range of micrograms or milligrams of material per cubic meter of air. For example, the OSHA permissible exposure limit (PEL) for graphite, our example material, is $5\,mg/m^3$. If the graphite is released into the air in the form of our 1 nm cubes, each of which weighs (from Table 3.1) $2.2 \times 10^{-24}\,kg$, we can calculate the corresponding number concentration, n:

$$n = (5\text{mg}/m^3)(10^{-6}\,\text{kg}/\text{mg})/2.2\times10^{-24}\,\text{kg}/\text{particle}$$
$$= 2.3\times10^{18}\,\text{particles}/m^3$$

This is an extremely high number concentration, which is physically impossible to create, meaning that the mass concentration of 1 nm graphite particles could never reach even a small fraction of the graphite PEL. In our research, and the measurements reported by others (Brouwer et al., 2009; Peters et al., 2009; Tsai et al., 2011), nanoparticle number concentrations rarely exceed 10^6 particles/cm³, or 10^{12} particles/m³. The mass concentration, m, corresponding to a 1 nm graphite number concentration of 1×10^{12} particles/m³ is:

$$m = (1\times10^{12}\,\text{particles}/m^3)(2.2\times10^{-24}\,\text{kg}/\text{particle})(10^6\,\text{mg}/\text{kg})$$
$$= 2.2\times10^{-6}\,\text{mg}/m^3 = 2.2\times10^{-3}\,\mu g/m^3 = 2.2\,\text{ng}/m^3$$

which is two million times lower than the graphite PEL. In addition, it is several thousand times smaller than the typical background mass concentration in the atmosphere, meaning it would be impossible to detect gravimetrically. The implications of such extremely low mass concentrations for nanoparticle sampling methodologies and exposure assessment will be discussed in detail in Chapter 7.

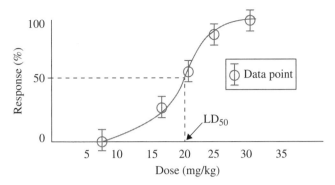

FIGURE 3.3   An idealized dose–response curve. mg/kg refers to the amount of the chemical in milligrams per kilogram of body weight of the subject. Graph courtesy of US National Institutes of Health.

3.4   DOSE METRICS: PARTICLE NUMBER, SURFACE AREA, MORPHOLOGY, AND SURFACE PROPERTIES

It is becoming increasingly evident that the toxicity of ENPs can be influenced strongly by factors beyond their size and material composition. As indicated in Table 3.1, the total aerosol surface area and particle surface-to-volume ratio both increase dramatically as particles move from the micrometer- to the nanometer-size range. For this reason, toxic effects that are more closely related to surface area than size may become more pronounced.

At this time, the importance of all of these properties is the subject of a great deal of research and has generated intense discussions among toxicologists. Toxicologists characterize the toxicity of various materials by way of dose–response curves (Fig. 3.3). Traditionally, dose is measured in terms of the mass of the material taken up by the test animal; for ENPs, however, there is general consensus that mass is not the correct metric, for reasons discussed above. The research and discussion now center on the basic question, that is, if mass is not the proper metric, what else is? Is it total particle number concentration? Total particle surface area? Over what range of sizes?

3.5   IMPLICATIONS FOR THE OCCUPATIONAL AND ENVIRONMENTAL HEALTH IMPACTS OF NANOPARTICLES

3.5.1   Respiratory Deposition

Nanoparticle deposition in the respiratory tract is discussed in some detail in Chapter 6; as indicated there, the respiratory deposition pattern for nanometer-sized particles is much different from that of larger particles. For example, only a relatively narrow range of micrometer-sized particles can penetrate to and deposit in the alveolar region, so that many particles in this size range will either be deposited in the upper airways and/or be inhaled and exhaled without depositing at all. Nanometer-sized particles can also be deposited in the alveolar region, raising the possibility that they can cause damage to lung tissue of the type described in Chapter 5 and/or penetrate the thin cell walls of the alveoli and surrounding capillaries and then be transported to other body systems.

In addition, the smallest nanoparticles deposit very efficiently in the nasal region, as do very large particles (aerodynamic diameter >50 μm). Large particles pose no particular problem (but rare materials such as nickel can cause nasal cancer), but it has been discovered that very small nanoparticles can deposit on the ends of the olfactory nerve; once collected, they can travel up the olfactory neurons to the brain. This serious exposure path is described more fully in Chapter 4.

3.5.2   Skin Penetration

In occupational health, it is commonly thought that intact adult skin (Fig. 3.4) presents an impervious barrier to particles. This is not the case for individual

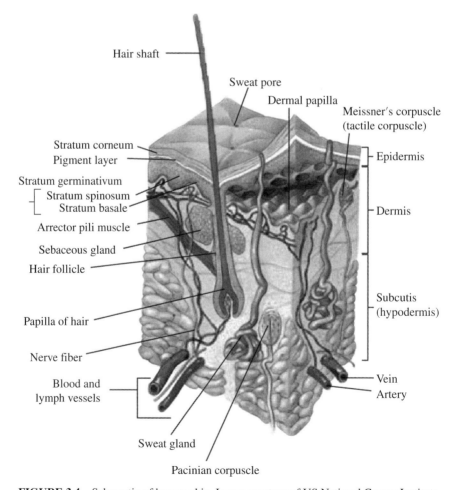

FIGURE 3.4   Schematic of human skin. Image courtesy of US National Cancer Institute.

molecules, many of which are known to penetrate the skin (Popendorf, 2006). Since very small nanoparticles approach molecular dimensions, the question naturally arises as to the potential for nanoparticles to behave more like a molecule and penetrate the skin. There is concern both for possible damage from the particles to the living layers of the skin and for the potential for nanoparticles to make their way to the circulatory system, where they could then travel to all other body organs. In addition, researchers are investigating the development of nanospheres for transdermal drug delivery, where the nanoparticles are *designed* to penetrate the skin (Desai et al., 2010).

At this time, there is little evidence that most ENPs penetrate intact skin, with the exception of those designed specifically to do so. Most recent studies have found no or limited skin penetration, but several studies have at least raised concerns for this mode of exposure. These subjects are discussed in more detail in Chapter 6.

3.6 IMPLICATIONS FOR PHYSICAL RISKS

3.6.1 Introduction

The only major physical risk from producing and using ENPs, beyond the normal risks associated with chemical manufacturing (slips and spills, electrical shock, etc.) is the risk of fire and explosion. This is a problem that is well understood within the pharmaceutical industry, which has been using micrometer-sized combustible powders for many years. The 2002 dust explosion at the West Pharmaceutical Plant in Kinston, NC (U.S Chemical Safety and Hazard Investigation Board, 2005), which killed 6 workers and injured 38 others, is a recent and unfortunate example of the hazards combustible dusts pose to the industry.

Since many ENPs are carbon based or consist of other combustible materials such as aluminum, and all ENPs by definition have a very high surface-to-volume ratio, the risk of fire and explosion when ENP aerosols are produced is both very real and not well understood or studied at this time. There are only a few references in the literature on this topic, which will be reviewed in the following sections.

3.6.2 Current Status

The Health and Safety Laboratory (HSL) of the Great Britain Health and Safety Executive (HSE) published a study which reviewed the state of the art for nanopowders as of 2004 (Pritchard, 2004); subsequently, they undertook an experimental program, the results of which were just recently published (Holbrow et al., 2010). Since that review article was published, several more studies have appeared in the literature. The Taiwan Institute for Occupational Safety and Health (IOSH) published the results of an investigation into a series of nanoaluminum explosions (Wu et al., 2010). The Nano Safe project of the European Union has conducted experimental explosivity measurements (Bouillard et al., 2008, 2009a, b, 2010).

Information on the explosion potential of ENPs takes three forms: theoretical predictions, experimental studies, and data from actual accidental explosions. Pritchard presents an excellent review of the theory and experimental investigations describing the behavior of dust explosions. Explosions for particles in the micrometer-size range have been studied extensively. The explosion of an organic dust consists of three distinct phases: pyrolysis or volatization, gas phase mixing, and gas phase combustion. Particle size primarily affects the first step in the process. Many experimental studies have found that the maximum particle size that can lead to an explosion is approximately 500 μm; presumably, particles larger than this do not pyrolize fast enough to sustain combustion. As particle size decreases (and surface area to volume increases), the minimum energy for ignition and the lower explosive limit (the minimum airborne concentration that will support combustion) decrease, while the strength of the explosion increases. For coal dust, these trends continue down to a particle size of approximately 50 μm, below which the maximum explosion pressure and rate of pressure rise level off. Although coal dust has been studied most extensively, the maximum pressure plateau for a few other materials has been reported. It is approximately 50 μm for flour, 40 μm for methylcellulose, and 10 μm for natural organic products such as starch and protein.

These results can be explained by considering the relative importance of the three phases of combustion described earlier. For coal dust in the particle size range from 50 to 500 μm, the particle pyrolysis is the rate-limiting step; below 50 μm, gas-phase mixing and/or combustion becomes the rate-limiting step. This analysis suggests that the explosion risk of nanometer-sized organic particles should be no more serious than the risk posed by such particles in the 10–50 μm size range.

Metal particles follow a different combustion sequence, consisting of melting, evaporation, and combustion. No size-limiting data were found in the literature, but the limited data available for metal particles suggest that the maximum pressure will plateau at considerably smaller diameters than for organic dusts. Experimental studies of the minimum energy required to initiate combustion (minimum ignition energy or MIE) show a very strong dependence on particle size, down to the smallest particles studied. Theoretical models predict that MIE is proportional to the cube of particle diameter, with no limiting plateau at smaller diameters. Experimental results for aluminum particles in the 0.1–100 μm size range are in agreement with the theory, with the MIE increasing approximately from 1 to 20 mJ (Bouillard et al., 2010). MIE varies considerably for different types of nanoparticles; published values range from less than 1 mJ for aluminum and iron nanopowders to greater than 1000 mJ for CNTs and copper nanopowders (Holbrow et al., 2010).

3.6.3   Conclusions

The limited testing and theoretical studies described above lead to the following general conclusions regarding the explosion risk of ENPs:

- Many ENPs, including carbon-based particle, metal particles, and polymer resins, are combustible
- In the micrometer-size range, the explosion risk is known to increase as the particles decrease in size

- This trend likely continues through the nanometer size range, but the explosion hazard potential will always be less than for the same material in a gas phase
- Most real NP aerosols consist of agglomerated particles, which will tend to lower the explosion hazard potential.

3.7 SUMMARY

ENPs, due solely to their very small size, present unique occupational and environmental health challenges. This has forced the occupational and environmental health specialist to take a fresh look at how the adverse effects of ENPs are evaluated and controlled. The current approach, as described in the remainder of this book, builds on what we have learned over many years of the properties of other aerosols, including micrometer-sized particles and natural and industrial nanoparticles.

REFERENCES

Bouillard J, Crossley A, Dien JM, Dobson P, Klepping T, Vignes A. What about explosivity and flammability of nanopowders? Brussels: NanoSafe, European Union; 2008.

Bouillard J, Vignes A, Dufaud O, Perrin L, Thomas D. Safety aspects of reactive nanopowders. In: Proceedings of the Annual Meeting AICHE. Nashville, TN; 2009a.

Bouillard J, Vignes A, Dufaud O, Perrin L, Thomas D. Explosion risks from nanomaterials. J Phys Conf Ser 2009b;170:012032. DOI: 10.1088/1742-6596/170/1/012032.

Bouillard J, Vignes A, Dufaud O, Perrin L, Thomas D. Ignition and explosion risks of nanopowders. J Hazard Mater 2010;15:873–880.

Brouwer DH, van Duuren-Stuurman B, Berges M, Jankowska E, Bard D, Mark D. From workplace air measurement results toward estimates of exposure? Development of a strategy to assess exposure to manufactured nano-objects. J Nanopart Res 2009;11:1867–1881.

Desai P, Patiolla RR, Singh M. Interaction of nanoparticles and cell-penetrating peptides with skin for transdermal drug delivery. Mol Membr Biol 2010;7:247–259.

Holbrow P, Wall M, Sanderson E, Bennett D, Rattigan W, Bettis R, Gregory D. Fire and explosion properties of nanopowders. Great Britain: Health and Safety Executive. Report No. RR782; 2010.

Jiang W, Mashayekhi H, Xing B. Bacterial toxicity comparison between nano- and micro-scaled oxide particles. Environ Pollut 2009;157:1619–1625.

Peters TM, Elzey S, Johnson R, Park H, Grassian VH, Maher T, O'Shaughnessy P. Airborne monitoring to distinguish engineered nanomaterials from incidental particles for environmental health and safety. J Occup Environ Hyg 2009;6:73–81.

Popendorf W. *Industrial hygiene control of airborne chemical hazards*. Boca Raton (FL): Taylor & Francis; 2006.

Pritchard DK. Literature review—explosion hazards associated with nanopowders. Great Britain: Health and Safety Executive. Report No. HSL/2004/12; 2004.

Tsai CJ, Huang CY, Chen SC, Ho CE, Huang CH, Chen CW, Chang CP, Tsai SJ, Ellenbecker MJ. Exposure assessment of nano-sized and respirable particles at different workplaces. J Nanopart Res 2011;13:4161–4172.

U.S. Chemical Safety and Hazard Investigation Board. 2005. Dust explosion at West Pharmaceutical Services. Washington (DC): U.S Chemical Safety and Hazard Investigation Board.

Wu HC, Ou HJ, Hsiao HC, Shih TS. Explosion characteristics of aluminum nanopowders. Aerosol Air Qual Res 2010;10:38–42.

4

ROUTES OF EXPOSURE FOR ENGINEERED NANOPARTICLES

4.1 INTRODUCTION

Current knowledge on the toxicity of engineered nanoparticles will be discussed in Chapter 5. The fact that a substance has the *potential* to cause damage does not necessarily mean, however, that actual damage will occur. This requires that a person be *exposed* to the toxic material, and that the material enters the body as a result of that exposure and travels to a body organ where damage can occur. Thus, in evaluating the potential hazards of engineered nanoparticles (ENPs), *exposure characterization* and ultimately *exposure assessment* are as equally important as the study of biological effects.

Since exposure characterization and exposure assessment are key concepts in this book, it is important to start with definitions. Beginning with exposure assessment, Rappaport and Kupper (2008) state the following:

> The term *exposure* implies contact between a chemical contaminant and a portal of entry into the body, i.e., the lungs, gastrointestinal tract, or the skin. Exposure can be quantified either as the level of the contaminant at the point of contact (e.g., mg/m^3 of air in the breathing zone) or as the level of a contaminant (or its products) inside the body (e.g., μg/l of blood or urine). Thus, the term *exposure assessment* refers to the estimation of parameters characterizing the distributions of environmental and/or biological levels of a contaminant (or its products) across an exposed population, along with attendant statistical evaluations and interpretations of such parameter estimates.

Exposure Assessment and Safety Considerations for Working with Engineered Nanoparticles, First Edition. Michael J. Ellenbecker and Candace Su-Jung Tsai.

As will become evident later in this book, the health and safety field is long way from actually performing exposure assessments, as defined above, for most ENP exposure scenarios. At best, we can use available instruments and techniques to measure the exposures at certain locations and times, which may be representative of a very limited actual exposure scenario. For example, a workplace exposure assessment requires the workers being evaluated to actually wear the sampling equipment, so that a personal breathing zone sample can be collected; as discussed in Chapter 7, such equipment is far from being available for evaluating ENP exposures. In this book, we have chosen to characterize currently available ENP measurement methods as *exposure characterization*; this implies a much more limited exposure evaluation strategy than exposure assessment and is consistent with the Environmental Protection Agency's definition: "exposure characterization is the risk analysis step in which human interaction with an environmental agent of concern is evaluated" (EPA, 2012).

As indicated above, there are three primary routes of exposure for toxic chemicals, including ENPs, that is, inhalation, dermal penetration, and ingestion. It is generally thought that the most important route of exposure is inhalation, followed by dermal and ingestion.

4.2 ENGINEERED NANOPARTICLE EXPOSURE THROUGH INHALATION

The occupational and environmental health community has studied exposure to aerosols by inhalation for many years, so there is a large knowledge base that can be applied to ENPs. The primary issues that must be addressed are the regional deposition of ENPs in the respiratory tract and the fate of those particles once they are deposited. Respiratory deposition is a combined function of the aerosol properties of the nanoparticle and the structure and function of the respiratory system, while particle fate is closely related to the toxicology of the particle.

4.2.1 Human Respiratory System

A schematic of the human respiratory system is shown in Figure 4.1. This system evolved over millions of years to fill its primary function of gas exchange, where oxygen is taken in and carbon dioxide is expelled. If we follow the respiratory tract from its inlet at the nose or mouth, we pass through the trachea and enter the branching system of bronchi. After approximately 22 levels of branching, we enter the terminal bronchioles and finally the alveoli, where gas exchange takes place. The pulmonary circulatory system follows a similar branching pattern, ending with capillaries that surround each alveolus. Oxygen diffuses through the alveolar and capillary cell walls, where it is picked up by the red blood cells, and carbon dioxide diffuses in the opposite direction and is exhaled.

The body also evolved methods to protect itself against particles that might deposit in the respiratory system, in particular naturally occurring viable and dangerous aerosols such as bacteria and viruses. The airways of the bronchial tree

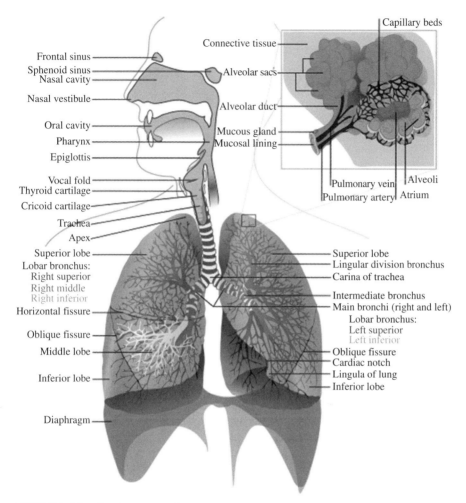

FIGURE 4.1   The human respiratory system. Image used by permission, Wikimedia/LadyofHats.

are lined with cilia and mucous-secreting cells, which comprise the so-called mucociliary escalator. The cilia beat back and forth and move the mucus toward the top of the trachea, where it is swallowed. The fate of particles that deposit in the mucus depends on the nature of the particle. Liquid- and water-soluble solid particles will dissolve in the mucus and can diffuse through the bronchial wall and enter the circulatory system as individual molecules. This is the principle by which respiratory drug-delivery systems operate. Nonsoluble solid particles will imbed in the mucus and travel up the escalator, where they are swallowed and cleared from the body through the digestive system.

The alveoli are not lined with cilia and mucus cells, since they would interfere with gas exchange. Fortunately, the body has evolved a second line of defense, namely, alveolar macrophages. These are phagocytic cells that move freely on the surfactant film coating the alveolus, with the mission to protect the alveolus from bioaerosols such as bacteria and viruses. When they encounter a deposited particle, they attempt to engulf it; once taken into the macrophage, enzymes attack the bacterium or virus and kill it. If the macrophage encounters an inert particle such as coal dust, it will engulf it and carry it with it as it moves across the alveolar surface. Eventually, the macrophage moves randomly to the bottom of the mucociliary escalator and is cleared, along with any engulfed particles.

Alveolar macrophages evolved over millions of years to protect the alveolus against naturally occurring aerosol particles; unfortunately, they respond poorly when encountering some aerosols related to our industrial society. As described above, macrophages can engulf inert particles such as coal dust and clear them to the mucociliary escalator. Coal miners, however, can develop coal miners' pneumoconiosis, or black lung; this disease is caused simply by an overwhelming of the macrophage clearance method due to massive numbers of coal dust particles being deposited in the alveoli. Other industrial materials are not so benign. Crystalline silica is toxic to macrophages; when such a particle is engulfed, the surface chemistry is such that the macrophage's membrane is disrupted, the cell bursts, and the enzymes are released onto the surface of the alveolus, causing a fibrotic reaction that leads to the progressive occupational disease silicosis. It is a progressive disease in that one silica particle can kill many macrophages, and thus the disease can worsen even after exposure ends. Asbestos fibers cause a similar disease, asbestosis, by a somewhat different mechanism. The fibers of concern are too long to be engulfed by the macrophage (Fig. 4.2), which cause the macrophage to burst with the same effects as silica. As discussed in Chapter 5, similar fibrotic effects have been found for carbon nanotubes, indicating that they are also lethal to macrophages.

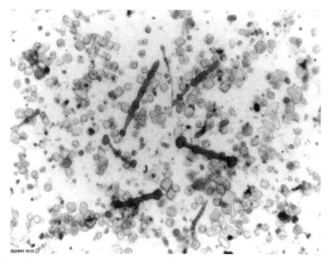

FIGURE 4.2 Photomicrograph of asbestos fibers deposited in a lung alveolus. The lung tissue has been digested. Note the barbell-shaped remains of alveolar macrophages on the fibers. Image used by permission Wikimedia/Ospedale S. Polo.

4.2.2 Particle Deposition in the Respiratory System

In order to determine the potential for ENPs to cause harm, it is not enough to assess their likelihood of depositing in the human respiratory system; it is also important to know exactly *where* in the respiratory system they are likely to deposit. As discussed above, the respiratory system has evolved to include the mucociliary escalator, which is quite efficient at clearing particles that deposit in it. Consequently, we are primarily concerned about particles that may deposit at both extremes of the respiratory system, that is, the nasal region and the alveoli. As discussed in Chapter 5, ENP deposition on the olfactory nerve ends presents a possible direct path to the central nervous system. With alveolar deposition, we are concerned both with direct damage to the alveolar tissue (inflammation, fibrosis, etc.) but also with the potential for ENPs to penetrate the very thin cellular walls where they subsequently can be carried by the circulatory system to all of the other organs. For CNTs, there is also concern for their penetration into the pleural space, leading possibly to the development of mesothelioma.

There are several variables that affect particle deposition patterns in the lung. The most important is breathing pattern, which is affected by exertion level. A sedentary individual typically breathes through the nose at a low breath rate and volume. As the work level increases, breathing switches to the mouth and breathing becomes more frequent with each breath taking in more air. A second variable is the person's size; a child will have a different pattern than a petite woman that will be different from a large man. Occupational hygienists usually assume that we can limit our model to an "average" adult working at a moderate level of exertion. The danger from olfactory deposition, however, may be greatest for more moderate levels of exertion, where nasal breathing predominates; thus in a somewhat perverse way, workers at hard labor may be at least partially protected from this respiratory danger.

Our knowledge of the regional deposition patterns of nanoparticles is based on a combination of experimental studies and theoretical modeling. Experimental studies are limited by our inability to measure particles in regions of living human lungs and the differences in structure between human lungs and those of experimental animals (e.g., the quite different airway dimensions in human and rat lungs). Because of this, most of our assumptions about the regional deposition of nanoparticles are based on theoretical models. The most widely accepted and used model is that developed by the International Commission on Radiation Protection (ICRP) (1994), which has also been adopted by the American Conference of Governmental Industrial Hygienists (ACGIH) (2012). Figure 4.3 shows the ICRP inhalation deposition model and predicts that significant numbers of NPs can deposit in both the nasal and alveolar regions.

In occupational hygiene, particles that deposit in the alveolar region are called *respirable* particles. As seen in Figure 4.3, the alveolar deposition curve is bimodal, with peaks centered at about 20 nm and 2 μm. Historically, it has been the micrometer-sized particles that were considered as respirable, and procedures have been developed to measure them gravimetrically. This respirable mass sampling is performed in two stages, that is, first a pre-separator separates the respirable from the nonrespirable particles and next the respirable particles are collected on a preweighed filter for

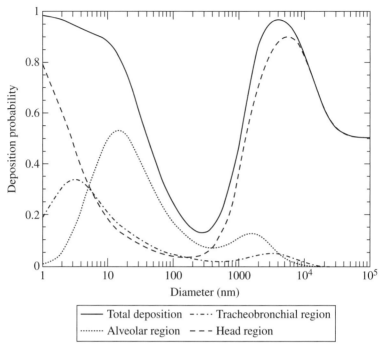

FIGURE 4.3 Modeled total particle deposition probability in the respiratory tract and deposition probability in the various lung regions (ICRP, 1994). Deposition has been modeled assuming an adult breathing through their nose at 25 l/min (light exercise) and exposed to spherical particles with a density of 1000 kg/m³. Reprinted with permission from ICRP (1994). Copyright 1994 Elsevier, Inc.

gravimetric analysis. The ACGIH respirable mass sampling curve, shown in Figure 4.4, has been adapted from the ICRP model for mouth-breathing. The most common device for meeting the pre-separator curve is a small personal cyclone, such as the one shown in Figure 4.5; single-stage impactors have also been widely used.

For our purposes, it is important to recognize that the standard respirable mass sampling method described above will not work for detecting respirable nanoparticles, that is, the particles making up the left peak in Figure 4.3. The reason for this is simple—these particles have so little mass that gravimetric sampling has little chance of detecting a measurable mass, unless a very large volume of air is sampled. If respirable particles are present in both the nanometer and micrometer size ranges, the mass of the micrometer particles will completely overwhelm that of the nanometer-sized particles, since each 2 μm particle has ten thousand times the mass of each 20 nm particle.

Once ENPs deposit in the alveolar region, there is concern both for direct damage to the alveoli (i.e., inflammation and fibrosis—see Chapter 5) and for the possible penetration of NPs through the thin alveolar cell walls; suggested mechanisms for alveolar penetration include "…endocytosis, transcytosis, or unidentified cellular

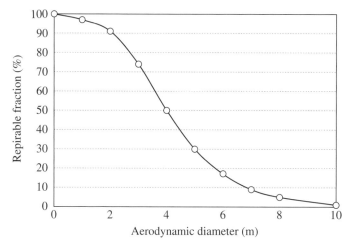

FIGURE 4.4 Respirable fraction as a function of particle aerodynamic diameter. Adapted from ACGIH respirability criteria.

FIGURE 4.5 Personal cyclone, used for respirable mass sampling. Photo courtesy of BGI, Inc.

mechanisms" (Nel et al., 2006). Once nanoparticles penetrate the alveolar cell wall, they can then penetrate the adjacent capillary cell wall and be transported to all other body systems. Nel suggests that resulting systemic effects can be assessed using "…assays for oxidative stress (e.g., lipid peroxidation), C-reactive protein, immune and inflammatory responses, and cytotoxicity (e.g., release of liver enzymes and glial fibrillary acidic protein)" (Nel et al., 2006).

Figure 4.3 demonstrates that very small ENPs will deposit preferentially in the nasal region. If they deposit at random locations, this may not present a hazard, but unfortunately there is evidence that these very small NPs can be taken up by the olfactory nerve ends and be transported up the nerve to the brain (Elder et al., 2006; Oberdörster et al., 2004). This important new route to central nervous system toxicity is discussed in more detail in Chapter 5.

4.3   ENGINEERED NANOPARTICLE EXPOSURE THROUGH DERMAL CONTACT

While occupational hygiene has long been concerned about the potential for various toxic liquids and gases to penetrate the skin, it has always been assumed that intact skin presents an impermeable barrier to particles. As particle size has been reduced to the NP range, however, concern grew as to the potential for these very small particles to penetrate the skin and then be transported via the circulatory system to other body organs. Consequently, a great deal of work has investigated this topic in recent years (Crosera et al., 2009; Larese Filone et al., 2009; Nohynek et al., 2008; Rouse et al., 2007; Ryman-Rasmussen et al., 2006; Zhang et al., 2008). The conclusions of Rouse et al. seem to summarize the current state of knowledge concerning the skin penetration of fullerenes, among the smallest ENPs:

> This study confirms that fullerene-based peptides can penetrate intact skin and that mechanical stressors, such as those associated with a repetitive flexing motion, increase the rate at which these particles traverse into the dermis. These results are important for identifying external factors that increase the risks associated with NP exposure during manufacturing or consumer processes. Future assessments of NP safety should recognize and take into account the effect that repetitive motion and mechanical stressors have on NP interactions with the biological environment. Additionally, these results could have profound implications for the development of NP use in drug delivery, specifically in understanding mechanisms by which NPs penetrate intact skin.

It might be expected that quantum dots' very small size might enhance their skin penetration ability (Lin et al., 2008). Ryman-Rasmusson et al. (2006) reached the following conclusions for quantum dots:

> We selected commercially available quantum dots (QD) of two core/shell sizes and shapes and three different surface coatings to determine if QD could penetrate intact skin in a size- or coating-dependent manner. Spherical 4.6 nm core/shell diameter QD 565 and ellipsoid 12 nm (major axis) by 6 nm (minor axis) core/shell diameter QD 655

with neutral (polyethylene glycol), anionic (carboxylic acids) or cationic (polyethylene glycol-amine) coatings were topically applied to porcine skin in flow-through diffusion cells at an occupationally relevant dose for 8 h and 24 h. Confocal microscopy revealed that spherical QD 565 of each surface coating penetrated the stratum corneum and localized within the epidermal and dermal layers by 8 h. Similarly, polyethylene glycol– and polyethylene glycol-amine–coated ellipsoid QD 655 localized within the epidermal layers by 8 h. No penetration of carboxylic acid–coated QD 655 was evident until 24 h, at which time localization in the epidermal layers was observed. This study showed that quantum dots of different sizes, shapes, and surface coatings can penetrate intact skin at an occupationally relevant dose within the span of an average-length work day. These results suggest that skin is surprisingly permeable to nanomaterials with diverse physicochemical properties and may serve as a portal of entry for localized, and possibly systemic, exposure of humans to QD and other engineered nanoscale materials.

On the other hand, Zhang et al. (2008) suspended QDs consisting of a cadmium/selenide core with a cadmium sulfide shell coated with polyethylene glycol (PEG) in water and measured their penetration through porcine in flow-through diffusion cells. They characterized the penetration into skin as "…minimal and limited to the uppermost SC [stratum corneum] layers and areas near hair follicles."

Larese Filone et al. (2009) applied 25 nm Ag particles to intact and damaged human abdominal skin in a Franz cell, an *in vitro* skin permeation device, and found particle penetration in each case, with the penetration through damaged skin approximately five times that of intact skin. They conclude that "…rigid NPs smaller than 30 nm are able to penetrate passively the skin, reaching the deepest layers of the stratum corneum and the outermost surface of the epidermis … there is an appreciable increase in permeation using damaged skin."

Of particular concern is the possible dermal penetration of ZnO and TiO_2 nanoparticles, since they are both used in sunscreens. Crosera et al. (2009) reviewed the research on this topic, including the large 3-year European project NANODERM which "…involved a great number of research groups in the evaluation of skin permeation of different TiO_2-based sunscreens. The project provided many data from *in vivo* and *in vitro* experiments with human and porcine skin, with human foreskin transplanted to immunodeficient mice, and with dermal cells in culture." They report conflicting results from the various research groups; some reported a small amount of penetration into the epidermal layer, while others found no penetration through the stratum corneum. They conclude that "the project report confirmed the safety of the sunscreens formulation containing TiO_2 nanoparticles, reporting no evidence of nanoparticle transcutaneous penetration."

Other studies have reached somewhat different conclusions. Bennat and Müller-Goymann (2000) applied formulations containing "microfine" ($d_p \approx 30$ nm) TiO_2 to human skin samples and found that it penetrated deeper into the skin when an oily dispersion was used compared to an aqueous one. They suggest that this is caused by the absorption of the oily carrier by lipid cells which enhances particle transport. In any case, penetration was found only into the stratum corneum, although they state that "penetration into deeper layers of the skin … seems to be possible and should be examined further."

Taken together the literature on NP dermal exposure seems to indicate that (i) the smaller the NP, the more likely it is to penetrate the skin and (ii) damaged skin presents a ready pathway for NP penetration.

4.4 ENGINEERED NANOPARTICLE EXPOSURE THROUGH INGESTION

Ingestion of toxic materials is rare in today's industrial environment, although it was common early in the Industrial Revolution. The classic pathway is contamination of the hands with the material, followed by hand-to-mouth transfer, typically from handling food or cigarettes. With the widespread recognition of the need for good hygiene on the shop floor, this should be a relatively rare occurrence today. In addition, as discussed in Chapter 3, the mass of each individual nanoparticle is very small. Since the known toxic endpoints from ingestion are largely related to mass, the combination of low probability of occurrence and low mass dose if exposure does occur should limit our concern for this route of exposure. The evidence for toxicity from ENP ingestion is summarized in Section 5.3.

4.5 TRANSLOCATION OF NANOPARTICLES FROM THE LUNG

Kreyling et al. (2010), in a commentary accompanying an article by Choi et al. (2010), state:

> Little is known about the fate of nanoparticles that enter the lungs, either deliberately through medical treatments or incidentally through air pollution and occupational exposure in the workplace … Translocation and accumulation in tissues seem to depend on physicochemical properties of the nanoparticle core and surface. Thus far, however, few studies have described nanoparticle transport across the lungs quantitatively.

Since NPs depositing in the airways are likely to be cleared via the mucociliary escalator, the primary concern is the fate of particles that deposit in the alveoli. In order to translocate from the alveolar surface to the circulatory system, an NP must first avoid capture by an alveolar macrophage, then pass through the fluid lining the alveolar surface, then the alveolar epithelial cells, followed by the endothelial cells of the capillary wall. Once they enter the capillary, of course, they can be carried via the circulatory system to all other body organs.

Choi et al. (2010) studied the translocation of a range of ENPs following instillation into rat lungs. They used nine "inorganic/organic hybrid nanoparticles (INPs)" consisting of a two-layer quantum dot and an organic coating, and seven "organic nanoparticles (ONPs)" of varying composition. The INPs ranged from 5 to 320 nm in diameter and the ONPs from 7 to 220 nm. A variety of surface charge patterns were present. They found that particles smaller than 38 nm (INP) or 34–48 nm (ONP) translocated rapidly (<30 min) from the alveoli to lymph nodes, but larger NPs did not.

Among the NPs smaller than the threshold, they found that surface charge played a critical role for rapid translocation. Hydrophilic and neutral surface charges led to rapid translocation, while NPs with highly charged surfaces were captured either in the alveolar fluid or epithelial cells and did not translocate. The smallest NPs (5 nm diameter) migrated to the lymph nodes within 3 min and to the kidneys and urine within 30 min. They conclude that surface-charge neutral NPs with diameters <34 nm "…are potentially the most dangerous in that they rapidly traverse the epithelial barrier, enter regional LNs [lymph nodes], and if ≥6 nm are not cleared efficiently by the kidneys. Thus, exposure of internal organs and tissues will be maximal." Importantly, even for the smallest particles where translocation rapidly occurred, "the majority of the administered NPs in our study remained in the lungs, with small amounts rapidly appearing in the mediastinal LNs [lymph nodes] within 30 min post-administration."

These results are consistent with Möller et al. (2008), who had human volunteers inhale 100 nm carbon particles and found that most particles were retained in the lung "…without substantial systemic translocation or accumulation in the liver at 48 h." They are also consistent with Kreyling et al. (2002), who exposed rats to 15 and 80 nm diameter iridium particles by inhalation and found that "minute particle translocation of <1% of the deposited particles into secondary organs such as liver, spleen, heart, and brain was measured after systemic uptake from the lungs. The translocated fraction of the 80-nm particles was about an order of magnitude less than that of 15-nm particles." In a study with rats, the same team of researchers (Semmler et al., 2004) found that about 0.2% of 15–20 nm iridium particles translocated to each of the brain, liver, spleen, and kidney.

4.6 SUMMARY

Of the possible routes of exposure to ENPs, inhalation is by far the most likely. It is also the most likely to lead to adverse health effects, due to the ability of nanometer-sized particles to deposit in the alveoli and subsequently cause direct damage to the lung or to translocate to other body systems. Although dermal contact is also likely under certain circumstances, the evidence for ENP penetration through healthy adult skin is conflicting at this time; the greatest concern should probably be for products, such as sunscreens, that are meant to be applied directly to the skin, and for exposure to damaged skin. The last route of exposure, ingestion, should be relatively easy to avoid.

REFERENCES

American Conference of Governmental Industrial Hygienists [ACGIH]. *2012 TLVs and BEIs—Based on the Documentation of the Threshold Limit Values for Chemical Substances and Physical Agents & Biological Exposure Indices*. Cincinnati (OH): ACGIH; 2012.

Bennat C, Muller-Goymann CC. Skin penetration and stabilization of formulations containing microfine titanium dioxide as physical UV filter. Int J Cosmet Sci 2000;22:271–283.

Choi HS, Ashitate Y, Lee JH, Kim SH, Matsui A, Insin N, Bawendi MG, Semmler-Behnke M, Frangioni JV, Tsuda A. Rapid translocation of nanoparticles from the lung airspaces to the body. Nat Biotechnol 2010;28:1300–1303.

Crosera M, Bovenzi M, Maina G, Adami G, Zanette C, Florio C, Larese FF. Nanoparticle dermal absorption and toxicity: a review of the literature. Int Arch Occup Environ Health 2009;82:1043–1055.

Elder A, Gelein R, Silva V, Feikert T, Opanashuk L, Carter J, Potter R. Translocation of inhaled ultrafine manganese oxide particles to the central nervous system. Environ Health Perspect 2006;114:1172–1178.

International Commission for Radiation Protection [ICRP]. Human respiratory tract model for radiological protection. Ann ICRP 24:1-482. Ottawa (CA): ICRP; 1994.

Kreyling WG, Hirn S, Schleh C. Nanoparticles in the lung. Nat Biotechnol 2010;28:1275–1276.

Kreyling WG, Semmler M, Erbe F, Mayer P, Takenaka S, Schulz H. Translocation of ultrafine insoluble iridium particles from lung epithelium to extrapulmonary organisms is size dependent but very low. J Toxicol Environ Health A 2002;65:1513–1530.

Larese Filone F, D'Agostina F, Croserab M, Adamib G, Renzic N, Bovenzia M, Maina G. Human skin penetration of silver nanoparticles through intact and damaged skin. Toxicology 2009;255:33–37.

Lin P, Chen JW, Chang LW, Wu JP, Redding L, Chang H, Yeh T, Yang CS, Tsai MH, Wang HJ, Kuo YC, Yang RSH. Computational and ultrastructural toxicology of a nanoparticle, quantum dot 705, in mice. Environ Sci Technol 2008;42:6264–6270.

Möller W, Felten K, Sommerer K, Scheuch G, Meyer G, Meyer P, Haussinger K, Kreyling WG. Deposition, retention, and translocation of ultrafine particles from the central airways and lung periphery. Am J Respir Crit Care Med 2008;177:426–432.

Nel A, Xia T, Mädler L, Li N. Toxic potential of materials at the nanolevel. Science 2006;311:622–627.

Nohynek GJ, Dufour EK, Roberts MS. Nanotechnology, cosmetics and the skin: is there a health risk? Skin Pharmacol Physiol 2008;21:136–149.

Oberdörster G, Sharp Z, Atudorei V, Elder A, Gelein R, Kreyling W, Cox C. Translocation of inhaled ultrafine particles to the brain. Inhal Toxicol 2004;16:437–445.

Rappaport SM, Kupper LL. *Quantitative Exposure Assessment*. El Cerrito (CA): Stephen Rappaport; 2008.

Rouse JG, Yang J, Ryman-Rasmussen JP, Barron AR, Monteiro-Riviere NA. Effects of mechanical flexion on the penetration of fullerene amino acid-derivatized peptide nanoparticles through skin. Nano Lett 2007;7:155–160.

Ryman-Rasmussen JP, Riviere JE, Monteiro-Riviere NA. Penetration of intact skin by quantum dots with diverse physicochemical properties. Toxicol Sci 2006;91:159–165.

Semmler M, Seitz J, Erbe F, Mayer P, Heyder J, Oberdörster G, Kreyling WG. Long-term clearance kinetics of inhaled ultrafine insoluble iridium particles from the rat lung, including transient translocation into secondary organs. Inhal Toxicol 2004;16:453–459.

U.S. Environmental Protection Agency [EPA]. 2012. *What is Exposure Science?* EPA. Available at http://epa.gov/ncct/expocast/what.html. Accessed April 30, 2013.

Zhang LW, Yu WW, Colvin VL, Monteiro-Riviere NA. Biological interactions of quantum dot nanoparticles in skin and in human epidermal keratinocytes. Toxicol Appl Pharm 2008;228:200–211.

5

CURRENT KNOWLEDGE ON THE TOXICITY OF NANOPARTICLES

5.1 INTRODUCTION

The recent rapid growth in research on the basic properties and uses of engineered nanoparticles has been accompanied by a growing concern for the possible toxicity and adverse environmental impact of these new particles. Such concerns have found their way into the popular press. For example, a December, 2008 article in the Fashion section of the *New York Times* (Singer, 2008) titled "Skin deep—new products bring side effect: nanophobia" reviewed concerns over the possible toxicity of engineered nanoparticles used in cosmetics and sunscreens. She began with the following paragraph:

> It sounds like a plot straight out of a science-fiction novel by Michael Crichton. Toiletry companies formulate new cutting-edge creams and lotions that contain tiny components designed to work more effectively. But those minuscule building blocks have an unexpected drawback: the ability to penetrate the skin, swarm through the body and overwhelm organs like the liver.

The reference to Michael Crichton was not random. His 2002 novel, *Prey*, described a nanotechnology industry run amok, with self-replicating "nanorobots" escaping the laboratory and swarming through the air to lay waste to society. Fortunately for society, the laws of physics do not allow nanorobots to behave in this manner. As objects become smaller, viscous drag becomes much more important (see Chapter 2)—Crichton can

Exposure Assessment and Safety Considerations for Working with Engineered Nanoparticles,
First Edition. Michael J. Ellenbecker and Candace Su-Jung Tsai.

claim that the nanorobots move quickly by "climbing the viscosity," but a nanorobot flying through air would be very similar to a human swimming through very cold honey. In fact, assuming that such a device had sufficient wings and power, it could only move very slowly through air. As Freeman Dyson wrote in his review of *Prey* for the New York Review of Books (Dyson, 2003), "For nanorobots to behave like a swarm of insects, they would have to be as large as insects."

In spite of the physical impossibility of the nanoparticle actions presented in the book, *Prey* made a strong impression on the general public. Turning from fiction back to science, we can review the evidence for *real* concerns about the toxicity of ENPs.

5.2   THE TOXICITY OF INDUSTRIAL NANOPARTICLES

The creation of the Environmental Protection Agency (EPA) and the passage of the Clean Air Act Amendments in 1970 spurred a great deal of research on the health effects of hazardous air pollutants, including industrial nanoparticles—what EPA designates fine particles, or $PM_{1.0}$, particulate matter smaller than 1.0 μm in diameter. These industrial nanoparticles are produced primarily by combustion sources, such as diesel engines and coal-fired power plants, and from gas-to-particle conversion (e.g., smog formation) (Hinds, 1999). Several large-scale epidemiologic studies of the general population have documented the adverse health effects of fine particles; the first and most well-known is the so-called Harvard Six Cities study (Dockery et al., 1993). Starting in 1975, Dockery and his colleagues at the Harvard School of Public Health measured air pollution levels in six US cities, ranging from highly industrial locations (Steubenville, OH) to rural ones (Portage, WI) and evaluated mortality in those locations for the years 1979–1989. They found "…significant associations between mortality and inhalable, fine or sulfate particles," where they defined fine particle as those with an aerodynamic diameter less than 2.5 μm.

This study, and a similar one sponsored by the American Cancer Society, caused much consternation among various industries. Pope and Dockery (2006) reviewed the history of the various studies and discussions that followed, and concluded:

> Since 1997, there has been a substantial amount of research that added to the evidence that breathing combustion-related fine particulate air pollution is harmful to human health.

As indicated, most of the focus of these studies has been on $PM_{2.5}$, particles smaller than 2.5 μm, but more recent research has focused on particles in the nanometer-size range, called in the air pollution field "ultrafine" or "PM_1" particles. Regarding these particles, Pope and Dockery state:

> The most common indicator of fine PM is $PM_{2.5}$, consisting of particles with an aerodynamic diameter less than or equal to a 2.5-μm cut point (although some have argued that a better indicator of fine particles would be PM_1, particles with a diameter less than or equal to a 1-μm cut point).

5.3 NANOPARTICLE TOXICITY: GENERAL CONCEPTS

A recent review article (Ostrowski et al., 2009) used bibliographic techniques to review all of the ENP toxicology literature from 2000 to 2007. They found that approximately 900 toxicology articles had been published in almost 60 different journals since 2000. The articles were spread widely across the journals, with at most 18 articles published by a single journal, and the journals traversed many disciplines from pure toxicology to chemistry, biology, physics, and engineering. This trend can only have accelerated in the 6 years since the time period covered in Ostrowski's article.

This dispersion, combined with the large and rapidly growing *number* of ENPs, makes it much more difficult for any one individual or organization to stay current on the topic of nanoparticle toxicity. In this chapter, we will first discuss briefly the routes of exposure by which ENPs may enter the body in order to have a toxic effect, followed by summaries of current toxicity information for only some of the more common categories of ENPs.

5.3.1 Routes of Exposure

Broadly speaking, there are four routes by which toxic materials can enter the body: inhalation, skin penetration, ingestion, and injection. All of these are relevant for possible ENP exposures; injection differs from the other three, however, in that ENPs would be purposely injected to deliver therapeutic or diagnostic agents incorporated into the ENP. The other three exposure paths are relevant for *unintended* exposures to ENPs during their use.

Historically, inhalation has been by far the most common exposure route for toxins in the occupational environment, followed by skin penetration and then ingestion. These routes were described in some detail in Chapter 4 as they relate to human exposure. Here, we will only mention that either inhalation or injection may be used during animal testing to study possible adverse effects from inhalation, and inhalation can deliver the test dose either *chronically*, by inhalation over time, or *acutely*, where the entire dose is taken in by a single inhalation (called aspiration); this can complicate the interpretation of test results as applied to humans, as discussed in Section 5.4.

Ingestion, in addition to the purposeful swallowing of an engineered nanomaterials (ENM) (such as for drug delivery), can occur by hand-to-mouth transfer (such as from eating fruit with contaminated hands) or from particle clearance from the respiratory tract via the mucociliary escalator. Yah et al. (2012) recently reviewed the scientific literature concerning the fate of ENPs once they enter the digestive system, which documents that ENPs can both cause inflammatory responses in the gut and can be absorbed via lymph nodes and translocate to other bodily organs such as the liver, kidney, and spleen.

5.3.2 *In Vivo* and *In Vitro* Testing

Classically, toxicologists have assessed the toxicity of a particular material by administering it to live animals, typically (but not always) small rodents such as mice and rats. So-called *in vivo* toxicity testing, however, has many shortcomings—the tests are expensive, take a long time to complete, involve significant ethical issues, and the results must be translated from effects in the animal to effects in humans. These problems are compounded by the very large, and growing, number of ENMs for which toxicity data are needed. For example, determining the toxicity of C_{60} may seem relatively straightforward; but, in order to make fullerenes perform useful tasks, scientists *functionalize* them by adding chemical groups at one or more carbon site. Each one of these hundreds of different functionalized fullerenes is a new material, with the potential for unique toxicity. Since the number of new ENMs being developed exceeds the number of existing ENMs undergoing toxicity testing, science is falling further and further behind in the race to determine the hazards of nanotechnology (Maynard et al., 2006).

Recognizing this dilemma, many scientists are investigating the use of rapid *screening tests* to assess toxicity. Such *in vitro* testing alleviates most of the disadvantages of *in vivo* testing, except for the translation of the results to humans, which becomes even more difficult. Most scientists, however, believe that screening can fill the important role of determining the *relative toxicity* between ENMs, such as for example the wide variety of functionalized fullerenes or the carbon nanotubes (CNTs) produced by different manufacturers.

An even newer approach is to use computer models to predict response to materials; this approach, dubbed *in silico*, avoids the use of animals and cell cultures. The utility of the information from *in silico* modeling depends, of course, on the accuracy with which cellular response can be predicted. It is most successful when the molecule of concern is structurally similar to another molecule for which solid experimental toxicity data exist.

5.4 CARBON NANOTUBES

The toxicity testing of CNTs (Fig. 2.7) has received a great deal of attention in the last few years. Early studies tested CNTs in short-term single dose assays of pulmonary toxicity by instilling or aspirating CNTs into mice and rat lungs. In general, these studies found inflammation and granulomas comparable to the toxic dust quartz (Lam et al., 2004; Warheit et al., 2004). Shvedova et al. (2005, 2008), in both a single-dose aspiration study and a 4-day inhalation study, found an initial inflammatory response followed by granulomas, fibrosis, and decreased rates of respiration and bacterial clearance from the lungs. Similar results were reported by Ma-Hock et al. (2009).

Adverse effects have been found in other organs in addition to the respiratory system. Li et al. (2007b) found several markers of adverse effects on the cardiovascular system following aspiration by mice, indicating that the aspirated CNTs either penetrated the alveolar walls and traveled via the circulatory system to the heart, or caused secondary cardiovascular effects through mitochondrial oxidative perturbations in the lung. A recent paper (Legramante et al., 2012) found that "SWCNT pulmonary exposure might affect the cardiovascular autonomic regulation thus contributing to cardiac and arrhythmic events."

As mentioned in Section 3.2, there is a great deal of concern that CNTs may cause mesothelioma, a fatal cancer of the pleural lining of the lung that to date has only been associated with asbestos exposure. Two studies published in 2008 found asbestos-like effects in short-term bioassays when multiwalled CNTs (MWCNTs) were injected intraperitoneally into mice (Poland et al., 2008; Takagi et al., 2008), while a more recent inhalation study found that inhaled MWCNTs reached the subpleura of mice (Ryman-Rasmussen et al., 2009). In a follow-up study, Takagi (Takagi et al., 2012) used three different intraperitoneal doses of MWCNTs and found that, 1 year postinjection, the cumulative incidence of mesothelioma was dose dependent, ranging from 5/20 mice at the lowest dose to 19/20 at the highest dose. All surviving mice in the low-dose group "…had microscopic atypical mesothelial hyperplasia considered as a precursor of mesothelioma." An aspiration study conducted by NIOSH documented the actual penetration of MWCNTs from alveoli to the pleural space, as shown in Figure 5.1 (Mercer et al., 2010).

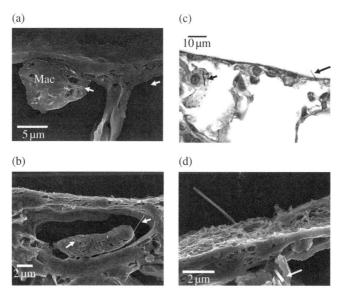

FIGURE 5.1   Representative micrographs of MWCNT in subpleural tissues, visceral pleura, and pleural space. In all four panels, the visceral pleural surface runs along the top of each micrograph. The FESEM image (a) shows a MWCNT loaded alveolar macrophage in an alveolus immediately beneath the visceral pleura surface. The right side of the image shows a single MWCNT fiber penetrating the alveolar epithelium into the subpleural tissues (80 µg dose, 28 day postaspiration). Image (b) shows a dilated subpleural lymphatic vessel which contains a mononuclear inflammatory cell that is penetrated by several MWCNT fibers (80 µg dose, 56 day postaspiration). A MWCNT penetrating the visceral pleura is shown in the light micrograph of image (c) with a MWCNT-loaded alveolar macrophage visible in the left side of the micrograph (80 µg dose, 28 day postaspiration). A single MWCNT penetrating from the subpleural tissue through the visceral pleura into the pleural space is shown in the FESEM image (d) (80 µg dose, 56 day postaspiration). Reprinted with permission from Mercer et al. (2010). Published under the BioMed Central Open Access license agreement.

There are many aspects of these first toxicology studies, beyond the usual difficulties inherent in any toxicology study of inferring human toxicity from animal data, which limit their direct applicability to human exposure. These issues are reviewed in some detail by Warheit et al. (2004). The term "carbon nanotube" does not define a single entity, in the way that "silica particle" does. There are different carbon nanotubes: single-walled and muli-walled, different catalysts, "dirty" versus purified, long versus short, and so on. Each toxicology study used either single-walled CNTs (Lam et al., 2004; Legramante et al., 2012; Li et al., 2007a; Mercer et al., 2010; Shvedova et al., 2005; Warheit et al., 2004) or MWCNTs (Ma-Hock et al., 2009; Mercer et al., 2010; Poland et al., 2008; Ryman-Rasmussen et al., 2009; Takagi et al., 2008), thus limiting the generalizability of the results. In addition, most of the published studies did not use CNT inhalation to deliver the dose; typically, a single dose was delivered via intratracheal instillation (Lam et al., 2004; Legramante et al., 2012; Warheit et al., 2004), aspiration (Li et al., 2007a; Mercer et al., 2010; Shvedova et al., 2005), or intraperitoneal injection (Poland et al., 2008; Takagi et al., 2008). The only studies that used inhalation were the ones carried out by Shvedova et al. (2008), Ma-Hock et al. (2009), and Ryman-Rasmusson et al. (2009); similar but more severe toxicological responses were found in the first two studies when inhalation was used rather than aspiration.

In order to maximize the possible response, the single doses used are extremely large compared to any possible occupational or environmental exposure. For example, Shvedova et al. (2005) exposed mice to single doses of single-walled CNTs via aspiration; the highest dose was 40 mg; assuming that a "typical" single-walled CNT can be modeled by a solid carbon cylinder 10 nm in diameter by 10 μm long, this is equivalent to 2×10^{12} CNTs. If a worker were exposed to 1 CNT/cm^3 (see exposure discussion below), (s)he would inhale approximately 10^7 CNTs in the course of one workday; the dose used in the NIOSH study thus would represent hundreds of years of occupational exposure.

In spite of the recognized shortcomings in the published CNT toxicology studies, there appears to be sufficient evidence for concern. Shvedova et al. (2005) exposed mice to the same doses of CNTs and silica, a known fibrinogen; the CNT-exposed mice developed fibrosis, while the silica-exposed mice did not. The studies that showed that MWCNTs can pass through the pleural membrane after inhalation and reach the subpleural wall and that found a possible association between CNT exposure and mesothelioma are of particular concern, since mesothelioma is a deadly disease associated historically with exposure to asbestos (Wagner et al., 1960). Because of the cancer risk, the OSHA 8-h time-weighted average permissible exposure limit for asbestos is only 0.1 fiber/cm^3 (OSHA, 2006). Controlling MWCNT exposures to a similarly low level will have a significant negative effect on the use by industry of this important type of nanoparticle.

5.5 FULLERENES

As described in Section 2.4, fullerenes are carbon molecules in a spherical configuration; the most common fullerene, C_{60}, has 60 carbon atoms (Fig. 2.8). The toxicity of fullerenes became a topic of great concern with the publication of a surprising report by Eva Oberdörster (2004). She had performed a fairly simple experiment, where she suspended C_{60} fullerene molecules in water, and then had largemouth bass swim in the water for 48 h, after which they were sacrificed. The surprise was that fullerenes were found in the brains of the bass. Under most circumstances, the brain is protected from direct attack by foreign molecules by the blood–brain barrier, but previous research had shown that nanometer-sized particles could travel to the brain by an alternative path, that is, uptake by the olfactory nerve ends in the nose and transport along the nerve to the brain. Oberdörster hypothesized that the fullerenes had reached the brain via this pathway, essentially taking a detour around the blood–brain barrier.

The fallout from the publication of this article was extensive, both in the popular press and in the scientific community. The popular press focused on the obvious implications for human toxicity, while the scientific community focused on Oberdörster's scientific methods. As several follow-on toxicity studies make clear, it is difficult to suspend fullerenes in water. For one thing, individual fullerenes are difficult to produce, since this molecule tends strongly to agglomerate. Thus, the studies are exposing animal populations to fullerene agglomerates (commonly called nC_{60}), and the size of the agglomerates can vary considerably from one study to another. In addition, in order to disperse the nC_{60} in water, solvents typically are used, meaning that there could be exposure to both the fullerenes and the solvents. These difficulties in generating a "standard" fullerene make it very difficult to compare toxicity studies done by different labs and to assess the actual risks from potential human exposures. The similarity to CNTs in this regard is striking.

Oberdörster et al. (2006) followed up their first study with a second one, where the fullerenes were carefully suspended in water by stirring for "at least" 2 months, which they characterized as an "environmentally relevant" solubilization protocol. At the maximum attainable concentrations of 35 ppm in fresh water and 22.5 ppm in seawater, they concluded:

> The results from the invertebrate studies show that an LC_{50}[1] could not be reliably calculated since nC_{60} concentrations high enough to cause 50% mortality could not be reached in the exposure media. This held true for three very different crustaceans: the freshwater daphnids and Hyalella, and the marine copepods. Therefore, acute toxicity of nC_{60} to these invertebrates is not a likely scenario in most expected environmental releases. However, significant sub-lethal effects—altered molting and decreased reproductive output—was found in daphnia [sic]. Although acute toxicity was not seen, these population-level effects should be taken into consideration when performing environmental risk assessments.

Following the Oberdörster experiments with fish, several studies attempted to duplicate this exposure pathway using inhalation. Gunter Oberdörster et al. (2004) exposed rats by inhalation to 36 nm diameter elemental carbon particles and found carbon particles in the olfactory bulb as early as one day postexposure. Elder et al. (2006) documented that when rats were exposed to 30 nm agglomerates of manganese oxide particles, they also traveled up the olfactory nerve to the olfactory bulb.

---

[1] A concentration in water that is lethal to 50% of the test animals.

Trpkovic et al. (2012), recently reviewed the literature concerning the toxicity of pristine and functionalized fullerenes. They conclude:

> The above-presented results indicate that, in addition to the protective antioxidant effects, both pristine and functionalized fullerenes display a range of activities that can cause cell death or dysfunction … Due to a relatively limited number of studies performed with each of the fullerene preparations, it is presently unrealistic to make definite conclusions about their toxicological behavior. However, it appears that most of the pristine and functionalized fullerene preparations are not overtly toxic unless photoexcited or used at very high concentrations that are unlikely to be encountered environmentally or during therapy.

Due to fullerene's very small size (diameter of about 0.7 nm), there has been considerable interest in their ability to penetrate the skin. Rouse et al. (2007) placed fullerene amino acid–derivatized peptide particles on porcine and flexed it. They found "…that fullerene-based peptides can penetrate intact skin and that mechanical stressors, such as those associated with a repetitive flexing motion, increase the rate at which these particles traverse into the dermis."

5.6   QUANTUM DOTS

Quantum dots (QDs) are used in a wide variety of applications in biology and medicine due to their extraordinary electrical and optical properties (Alivisatos, 1996; Gao et al., 2004; Howarth et al., 2005; Michalet et al., 2005). While many *in vitro* studies have shown QDs to be safe, there are still safety concerns regarding their use *in vivo*. Relatively recent reports suggest that QDs containing both Cd and Se are toxic and can damage certain types of cells, including normal as well as transformed or cancer cells (Gao et al., 2004; Li et al., 2009). Therefore, there is an urgent need for researchers to further investigate the toxicity factors of these QDs, especially in *in vitro* and *in vivo* biological systems, to better understand the potential for environmental and human health problems.

The toxicity of QDs is dependent on several factors, such as but not limited to surface coating, core composition, exposure time, size, and charge (Hoshino et al., 2004). The core of the CdSe and InP QDs usually have a shell or cap made with zinc sulfide (ZnS), which reduce the release of highly toxic metals and the generation of reactive oxygen species (Derfus et al., 2004). Unfortunately, the ion coatings themselves can lead to the increased toxicity in cells (Peng et al., 1997; Reiss et al., 2002).

5.7   METAL-BASED NANOPARTICLES

Of the many metal and metal oxide ENPs, the one that has received the most attention concerning its toxicity is nanometer-sized silver particles. Nanosilver is widely used as an antibacterial agent, which is a good indicator of its inherent toxicity. Holder and Marr (2013) recently reviewed the somewhat confusing and contradictory literature concerning the toxicity of silver nanoparticles and conclude:

> A safe level for airborne silver nanoparticles has yet to be determined. Inhaled silver has been detected in the blood, liver, brain, and kidneys of exposed rats. Despite the wide distribution of silver throughout the body, no adverse effects were observed in hematology and histopathology assessments at low doses ($\sim$0.06 mg m^{-3}). Animals exposed to silver subacutely at a high dose, 3.3 mg m^{-3}, showed minimal pulmonary inflammation or cytotoxicity. In contrast, animals exposed to a moderate dose, 0.5 mg m^{-3}, showed signs of chronic inflammation in the lungs and abnormalities in the liver. In vitro studies with silver nanoparticles have shown stronger effects, with many different cell lines showing reduced viability or oxidative stress response at doses ranging from the order of 1 μg mL^{-1} to 100 μg mL^{-1}. Cell studies have also shown a size-dependent effect; the smallest particles ($\sim$5–15 nm) required a lower mass dose to cause decreased viability and greater oxidative stress.

There are several possible explanations for the variation among *in vitro* studies and the differences between the *in vitro* and inhalation studies. Firstly, the properties of the silver nanoparticles used in each study likely differed. The inhalation studies were all performed with metallic silver nanoparticles (10–20 nm) condensed from silver vapor generated from either a spark discharge apparatus or a furnace. Alternatively, all of the *in vitro* studies were performed with silver nanoparticles either synthesized in solution or purchased in powder form, some of which had coatings, and resuspended in aqueous media. Secondly, the exposure route may have affected toxicity. Silver nanoparticles in cell culture media may aggregate into larger particles, obscuring the effects of the nanoparticles, or over time may release silver ions which can also cause a toxic effect apart from that of the nanoparticles.

Other commonly used metal-based ENPs include titanium dioxide (TiO_2), zinc oxide (ZnO), aluminum oxide (Al_2O_3), iron (Fe), and boron (B). In addition, elemental carbon, while not strictly speaking a metal, is used in nanoparticles in the form of more-or-less spherical particles of amorphous carbon. Research to date indicates that these materials, as a group, exhibit much less toxicity than the materials discussed above. The most common toxic response has been acute respiratory inflammation.

5.8   SUMMARY

This chapter only touched on the large and rapidly expanding literature concerning ENP toxicity. At this time, it is safe to say that most if not all of the most commonly used ENPs are toxic to some degree. Given the potential risks from exposure indicated by these studies, the prudent approach, in our opinion, is to take positive steps to evaluate and control such exposures. These are the topics of the remainder of this book.

REFERENCES

Alivisatos AP. Perspectives on the physical chemistry of semiconductor nanocrystals. J Phys Chem 1996;100:13226–13239.

Derfus AM, Chan WCW, Bhatia SN. Probing the cytotoxicity of semiconductor quantum dots. Nano Lett 2004;4:11–18.

Dockery DW, Pope CS, Xu X, Spengler JD, Ware JH, Fay ME, Ferris BG, Speizer FE. An association between air pollution and mortality in six U.S. cities. N Engl J Med 1993;329:1753–1759.

Dyson F. The future needs us. The New York Review of Books. New York; February 13, 2003.

Elder A, Gelein R, Silva V, Feikert T, Opanashuk L, Carter J, Potter R. Translocation of inhaled ultrafine manganese oxide particles to the central nervous system. Environ Health Perspect 2006;114:1172–1178.

Gao X, Cui Y, Levenson RM, Chung LWK, Nie S. In vivo cancer targeting and imaging with semiconductor quantum dots. Nat Biotechnol 2004;22:969–976.

Hinds WC. *Aerosol Technology—Properties, Behavior, and Measurement of Airborne Particles.* New York: Wiley-Interscience; 1999.

Holder AL, Marr LC. Toxicity of silver nanoparticles at the air-liquid interface. Biomed Res Int 2013;2013:328934.

Hoshino A, Fujioka K, Oku T, Suga M, Sasaki YF, Ohta T, Yasuhara M, Suzuki K, Yamamoto K. Physicochemical properties and cellular toxicity of nanocrystal quantum dots depend on their surface modification. Nano Lett 2004;4:2163–2169.

Howarth M, Takao K, Hayashi Y, Ting AY. Targeting quantum dots to surface proteins in living cells with biotin ligase. PNAS 2005;102:7583–7588.

Lam CW, James JT, McCluskey R, Hunter RL. Pulmonary toxicity of single-wall carbon nanotubes in mice 7 and 90 days after intratracheal instillation. Toxicol Sci 2004;77:126–134.

Legramante JM, Sacco S, Crobeddu P, Magrini A, Valentini F, Palleschi G, Pallante M, Balocchi R, Iavocoli I, Bergamaschi A, Galante A, Campagnolo L, Pietroiusti A. Changes in cardiac autonomic regulation after acute lung exposure to carbon nanotubes: implications for occupational exposure. J Nanomat 2012;2012:397206.

Li J, Li W, Xu J, Cai X, Liu R, Li Y, Zhao Q, Li Q. Comparative study of pathological lesions induced by multiwalled carbon nanotubes in lungs of mice by intratracheal instillation and inhalation. Environ Toxicol 2007a;22:415–421.

Li R, Dai H, Wheeler TM, Sayeeduddun M, Scardino PT, Frolov A, Ayala GE. Prognostic value of akt-1 in human prostate cancer: z computerized quantitative assessment with quantum dot technology. Clin Cancer Res 2009;15:3568–3573.

Li Z, Hulderman T, Salmen R, Chapman R, Leonard SS, Young SH, Shvedova AA, Luster MI, Simeonova PP. Cardiovascular effects of pulmonary exposure to single-wall carbon nanotubes. Environ Health Perspect 2007b;115:377–382.

Ma-Hock L, Treumann S, Strauss V, Brill S, Luizi F, Mertler M, Wiench K, Gamer AO, van Ravenzwaay B, Landsiedel R. Inhalation toxicity of multiwall carbon nanotubes exposed in rats for 3 months. Toxicol Sci 2009;112:468–481.

Maynard A, Aitken RJ, Butz T, Colvin V, Donaldson K, Oberdorster G, Philbert MA, Ryan J, Seaton A, Stone V, Tinkle S, Tran L, Walker N, Warheit DB. Safe handling of nanotechnology. Nature 2006;444:267–269.

Mercer RR, Hubbs AF, Scabilloni JF, Wang L, Battelli LA, Schwegler-Berry D, Castranova V, Porter DW. Distribution and persistence of pleural penetrations by multi-walled carbon nanotubes. Part Fibre Toxicol 2010;7:28.

Michalet X, Pinuaud FF, Bentolila LA, Tsay JM, Doose S, Li JJ, Sundaresan G, Wu AM, Gambhir SS, Weiss S. Quantum dots for living cells, in vivo imaging and diagnostics. Science 2005;307:538–544.

Oberdörster E. Manufactured nanomaterials (fullerenes, C_{60}) induce oxidative stress in the brain of juvenile largemouth bass. Environ Health Perspect 2004;112:1058–1062.

Oberdörster E, Zhu S, Blickley TM, McClellan-Green P, Haasch ML. Ecotoxicology of carbon-based engineered nanoparticles: effects of fullerene (C_{60}) on aquatic organisms. Carbon 2006;44:1112–1120.

Oberdörster G, Sharp Z, Atudorei V, Elder A, Gelein R, Kreyling W, Cox C. Translocation of inhaled ultrafine particles to the brain. Inhal Toxicol 2004;16:437–445.

Ostrowski AD, Martin T, Conti J, Hurt I, Herr Harthorn B. Nanotoxicology: characterizing the scientific literature, 2000–2007. J Nanopart Res 2009;11:251–257.

Peng X, Schlamp MC, Kadavanich AV, Alivisatos AP. Epitaxial growth of highly luminiscent CdSe/CdS core/shell nanocrystals with photostability and electronic accessibility. J Am Chem Soc 1997;119:7019–7029.

Poland CA, Duffin R, Kinloch I, Maynard A, Wallace WAH, Seaton A, Stone V, Brown S, MacNee W, Donaldson K. Carbon nanotubes introduced into the abdominal cavity of mice show asbestos-like pathogenicity in a pilot study. Nat Nanotechnol 2008;3:423–428.

Pope CA, Dockery DW. Health effects of fine particulate air pollution: lines that connect. J Air Waste Manage Assoc 2006;56:709–742.

Reiss P, Bleuse J, Pron A. Highly luminescent CdSe/ZnSe Core/Shell Nanocyrstals of low size dispersion. Nano Lett 2002;2:781–784.

Rouse JG, Yang J, Ryman-Rasmussen JP, Barron AR, Monteiro-Riviere NA. Effects of mechanical flexion on the penetration of fullerene amino acid-derivatized peptide nanoparticles through skin. Nano Lett 2007;7:155–160.

Ryman-Rasmussen JP, Cesta MF, Brody AR, Shipley-Phillips JK, Everitt JI, Tewksbury EW, Moss OR, Wong BA, Dodd DE, Andersen ME, Bonner JC. Inhaled carbon nanotubes reach the subpleural tissue in mice. Nat Nanotech 2009;4:747–751.

Shvedova AA, Kisin ER, Mercer R, Murray AR, Johnson VJ, Potapovich AI, Tyurina YY, Gorelik O, Arepalli S, Schwegler-Berry D, Hubbs AF, Antonini J, Evans DE, Ku B, Ramsey D, Maynard A, Kagan VE, Castranova V, Baron P. Unusual inflammatory and fibrogenic pulmonary responses to single-walled carbon nanotubes in mice. Am J Physiol Lung Cell Mol Physiol 2005;289:698–708.

Shvedova AA, Kisin ER, Murray AR, Johnson VJ, Gorelik O, Arepalli S, Hubbs AF, Mercer R, Keohavong P, Sussman N, Jin J, Yin J, Stone S, Chen BT, Deye G, Maynard A, Castranova V, Baron PA, Kagan VE. Inhalation vs. aspiration of single-walled carbon nanotubes in C57BL/6 mice: inflammation, fibrosis, oxidative stress, and mutagenesis. Am J Physiol Lung Cell Mol Physiol 2008;295:L552–L556.

Singer N. Skin deep—new products bring side effect: nanophobia. *New York Times*, December 4, 2008.

Takagi A, Hirose A, Nishimura T, Fukumori N, Ogata A, Ohashi N, Kitajima S, Kanno J. Induction of mesothelioma in p53+/- mouse by intraperitoneal application of multi-wall carbon nanotube. J Toxicol Sci 2008;33 (1):105–116.

Takagi A, Hirose A, Futakuchi M, Tsuda H, Kanno J. Dose-dependent mesothelioma induction by intraperitoneal administration of multi-wall carbon nanotubes in p53 heterozygous mice. Cancer Sci 2012;103 (8):1440–1444.

Trpkovic A, Todorovic-Markovic B, Trajkovic V. Toxicity of pristine versus functionalized fullerenes: mechanisms of cell damage and the role of oxidative stress. Arch Toxicol 2012;86:1809–1827.

U.S. Occupational Safety and Health Administration [OSHA]. Table Z-1 limits for air contaminants. Washington (DC): U.S. Occupational Safety and Health Administration; 2006. Report nr U.S. OSHA 29 CFR 1910:1000.

Wagner JC, Sleggs CA, Marchand P. Diffuse pleural mesothelioma and asbestos exposure in the North Western Cape province. Brit J Ind Med 1960;17:260–271.

Warheit DB, Laurence BR, Reed KL, Roach DH, Reynolds GAM, Webb TR. Comparative pulmonary toxicity assessment of single-wall carbon nanotubes in rats. Toxicol Sci 2004;77:117–125.

Yah CS, Simate G, Iyuke SE. Nanoparticles toxicity and their routes of exposures. Pak J Pharm Sci 2012;25:447–491.

6

SOURCES OF EXPOSURE

In order for an engineered nanoparticle (ENP) to cause an adverse health effect, two conditions must be met. First, the particle must be toxic, and second, of course, someone must be exposed to that particle. As discussed in Chapter 5, our knowledge of ENP toxicity is incomplete, but there is sufficient information to raise concern, at least for some types of nanoparticles. It is safe to say that we know somewhat more about potential sources of exposure. Although this book is concerned primarily with occupational exposures, we will briefly review other possible exposure scenarios.

6.1 OVERVIEW OF OCCUPATIONAL EXPOSURES

The rapid growth in the nanoparticle industry, as discussed in Chapter 1, means that the number of individuals with possible occupational exposures to ENPs is also growing rapidly. There are several ways to classify different occupational exposures. The first, of course, is by ENP *type*. In our experience, this has not proven to be particularly useful as a classification scheme, since exposures can occur for all types of ENPs, and the seriousness of the exposure is a stronger function of the factors discussed below than the particle type.

The second classification scheme is to distinguish between the manufacture of the nanoparticle itself and the *use* of that ENP in a product. Some companies both manufacture an ENP and then incorporate it into a finished product, but more

Exposure Assessment and Safety Considerations for Working with Engineered Nanoparticles,
First Edition. Michael J. Ellenbecker and Candace Su-Jung Tsai.
© 2015 John Wiley & Sons, Inc. Published 2015 by John Wiley & Sons, Inc.

commonly a company or facility does either one or the other. Exposure patterns during particle manufacture may be quite different than during device construction.

It is quite likely that exposures during the *research* phase for the development of a new nanoparticle or use will be different than during the *manufacturing* of that particle or product (scheme three). For example, during research only very small quantities of product typically are produced, compared to the manufacturing phase, but on the other hand, a research facility may have poorer engineering controls available to it.

A fourth scheme is to look at the *physical state* of the nanoparticle or the ENP-containing product. As a general rule, ENPs in dry powder form present the greatest potential for exposure, followed by ENPs suspended in a liquid, and the least exposure potential occurring for ENPs incorporated into a solid matrix. Workers, however, can be exposed to ENPs in each of these physical states, and the nature of the exposure varies considerably from powder to liquid to solid.

Once a nanoparticle or a product containing that particle reaches production, it is useful to consider possible occupational exposures in each phase of its life cycle, since the nature of those exposures varies considerably. Most simply stated, the major stages in a product life cycle are *manufacturing*, *use*, and end-of-life *disposal*. As a general rule, occupational exposures become less important as we move through a product's life cycle, and environmental exposures become more important.

Finally, we can categorize an exposure by its *route* into the body, that is, inhalation, dermal penetration, or ingestion. Again as a general rule, inhalation is usually the most important occupational exposure route, followed by dermal penetration, with ingestion a distant third.

The various possible ways to classify occupational exposures to ENPs make it difficult to organize this chapter in a logical fashion. Somewhat arbitrarily, we will consider research versus manufacturing first, followed by physical state, and finally life cycle state. Particle versus device manufacturing is considered within the manufacturing section below. Route of exposure is so fundamental that it is dealt with separately in the Chapter 7.

Only a few studies have been published that measured worker nanoparticle exposures in the industrial environment. Those studies are reviewed in detail in Chapter 8 and will be briefly discussed here when appropriate.

6.2   OCCUPATIONAL EXPOSURES IN RESEARCH FACILITIES

Typically, there are significant differences in possible exposures in research facilities when compared to manufacturing. Some of these differences tend to *reduce* exposure potential in research laboratories, while others *increase* the risk. The main factor tending to reduce exposure potential in research laboratories is the simple fact that research uses quantities of material that are orders of magnitude smaller than used in full-scale manufacturing. As a further complication, products typically move from research to *pilot-scale* production before full-scale production is started. In pilot-scale production, actual production equipment is used as opposed to laboratory equipment, but the *scale* of the equipment is much smaller than the full-scale production equipment. The quantities of material thus used in a pilot-scale operation typically fall somewhere between the quantities used in research and in full-scale production.

Another factor that sometimes reduces exposures in research facilities is the *energy* applied to nanoparticles. For example, a nanomaterial in dry powder form may be handled quite gently in the lab; once the process moves to manufacturing, the dry material may be moved mechanically at high velocity through the process, increasing the fraction of the material that is aerosolized by several orders of magnitude.

There are also factors that tend to increase exposures in research facilities. By their very nature, the equipment used in research operations changes constantly; this makes it more difficult to use certain engineering controls, such as isolation and ventilation, since these controls must be adapted to the equipment. One solution to this problem is to rely on general-duty ventilation controls, such as laboratory fume hoods or biological safety cabinets, when the scale of the research equipment allows it. This reliance may give the researchers a false sense of security, since it is assumed that such ventilation systems are 100% effective; research in our laboratory and by others indicates that this is not the case, as discussed in Section 9.1.

Contrary to the example given above, handling procedures may sometimes increase exposures in research facilities. Full-scale production may involve the use of completely mechanized processes, which isolate the material from the workers, whereas research may involve much more direct manipulation by the researcher.

Then there is the human factor to consider. Minimizing exposure to ENPs requires the use of good work practices, as discussed in Section 9.3. Workers in research laboratories may not be aware of the potential hazards of the materials they are working with, or, if they have been aware of those hazards, may tend to ignore them in the interests of making progress on their "science." In university laboratories, new students might join the laboratory group at any time, and these new students may not receive proper training on good work practices until they have been at work for some period of time. For these reasons, it is extremely important that the EHS staff responsible for research facilities have an aggressive program in place to ensure that all laboratory workers know the safe work practices they need to follow, and follow up to ensure that these practices are actually being implemented.

6.3 OCCUPATIONAL EXPOSURES IN MANUFACTURING FACILITIES

As discussed above, manufacturing facilities as a rule use larger-scale equipment than research facilities and process much larger quantities of material. These factors tend to increase the potential for exposure. On the other hand, production equipment may be more automated and enclosed than research equipment, and the use of local exhaust ventilation (LEV) might be easier since the equipment tends to stay in place for longer times. In addition, the work force at manufacturing facilities tends to have less turnover than at university research facilities, which rely on graduate students for much of their work.

Within manufacturing, we can expect that exposure patterns will be much different during the manufacturing of the ENP itself, as compared to the incorporation of the particle into a usable device. ENPs, like most other synthetic materials, are manufactured through a chemical synthesis process. Such processes, whether a chemical reaction (e.g., quantum dot manufacture) or a combustion process (e.g., carbon nanotube (CNT) manufacture), are done in a *closed* system. Thus, during the manufacturing process itself there are limited opportunities for exposure unless the enclosure is not complete. Combustion processes offer more opportunity for exposure than simple chemical reactions, since the products of combustion must be exhausted. Unless a suitable pollution control device is used (see Chapter 10), ENPs and other contaminants in the exhaust gas will go up the stack, where they may recirculate back into the workplace (Burgess et al., 2004) and/or lead to environmental exposures (Tsai et al., 2009a).

The greatest potential for exposure during ENP manufacture exists when the manufactured particles are harvested from the reaction vessel. At this point, the particles typically are in dry powder form, the closed system must be opened, and energy must be applied to the particles to transfer them from the reaction vessel to storage; this combination can easily lead to significant worker exposures (Maynard et al., 2004). In addition to production operations, reaction vessels typically must be cleaned periodically; such cleaning processes can lead to very high exposures, as Methner (2008) found and as discussed below.

It is more difficult to generalize about exposure potentials during the use of ENPs in manufacturing, since such manufacturing processes have great variability. For example, CNTs may be used as a reinforcing agent in a nanocomposite material; in this circumstance, the CNTs likely are added to an extruder in large quantities as a dry powder—a setting with the potential for very high exposures unless engineering controls are carefully used (Tsai et al., 2008a, 2009a). A very different use for the very same CNT might be as a memory element in an advanced memory storage device; here, the CNTS are manipulated one at a time in a liquid carrier, with very limited potential for exposure.

6.4   EXPOSURE POTENTIAL FOR ENPs IN DIFFERENT PHYSICAL STATES

ENPs can exist primarily in three basic physical states, with some specialized exceptions. The three primary states are as *dry powders*, *suspended in a liquid*, and *physically bound to a solid*. The term "physically bound to a solid" is meant to encompass a range of conditions where the ENPs are "trapped" by a solid, such as a nanocomposite where metal oxide ENPs are embedded in a polymer, ENPs in a coating on the surface of the material, and a digital memory chip, where CNTs are attached physically and electrically to a silica substrate. Specialized exceptions would include, for example, ENPs inside living cells, which are not "solids" or "liquids" in the sense meant here.

6.4.1   Dry Powders

Many ENMs are produced as dry powders during the manufacturing process. For example, Methner (2008) described a facility that was manufacturing metal and metal alloy nanospheres of approximately 15–50 nm diameter. The process used a "gas-phase condensation vapor deposition reactor" where micrometer-sized metal powders were vaporized and then condensed to nanometer-sized particles. The nanoparticles were scraped from the walls of the reactor and fell into a sealed container, where they were then taken away in dry powder form. CNTs are formed by similar processes. Measurements were first made without the use of LEV to control releases, and concentrations sometimes exceeded 100,000 particles/cm³ (after subtracting the background concentration); adding LEV, as discussed in Section 9.1, dramatically reduced exposures.

Obviously, such "nanopowders," once produced, must be handled and processed in order to produce useful products. Nanopowder handling presents the greatest potential for airborne exposure, since dry powders are easily aerosolized by relatively small amounts of energy. Experiments performed in our laboratory found that the simple task of transferring 100 g of nanoalumina from one beaker to another inside a fume hood (Fig. 6.1) can create particle concentrations *outside* the hood as high as 13,000 particles/cm³ (Tsai et al., 2009b). We have also found that more energetic operations can produce much higher airborne concentrations. For example, we have performed extensive evaluations of nanocompounding, where nanopowders are processed with polymer pellets in a twin-screw extruder to produce nanocomposites; these experiments are described in detail in Chapter 8. We found that the process of feeding nanoalumina from a hopper to the extruder released airborne nanoparticle concentrations as high as 1.3×10^6 particles/cm³ at the hopper and 2.8×10^5 particles/cm³ at the operator's workstation (Tsai et al., 2008b).

Besides the significant respiratory hazard presented by the handling of nanopowders, such handling can easily contaminate work surfaces and the handler's skin and clothing. This is illustrated by recent work in our laboratory using fluorescent-dyed nanoalumina particles and an ultraviolet light source. Figure 6.2 shows one of the hoods where we performed the particle transfer operation described above. Figure 6.2a shows the hood surface under visible light, where no particles are visible; under ultraviolet (UV) light

(a)   (b)

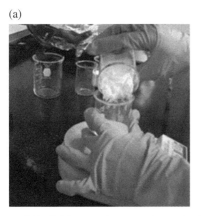

FIGURE 6.1 Pouring (a) and transferring (b) a nanopowder inside a fume hood. Photos by S. Tsai.

(a)   (b)

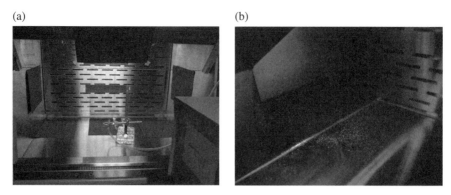

FIGURE 6.2 Enclosure used during nanoparticle experiment: (a) visible light and (b) ultraviolet light. Photos by S. Tsai.

(a)   (b)

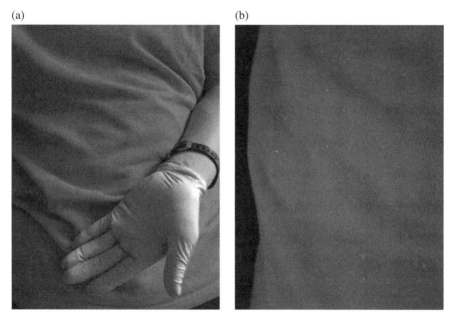

FIGURE 6.3 Nanoparticle contamination of a worker's shirt under (a) visible light and (b) UV light. Photos by S. Tsai.

(Fig. 6.2b), the surface contamination is very evident. Figure 6.3 shows the shirt worn by the student who performed the transfer operation. Again, under visible light, no contamination is seen (Fig. 6.3a), but the UV light uncovers many nanoparticles that collected on his shirt (Fig. 6.3b). Figure 6.3 reveals that the use of a particle-handling hood did not prevent the migration of nanoparticles from inside the hood to his clothing. (Note: our students routinely wear protective clothing such as disposable laboratory coats when performing experiments; here, wearing only a shirt was done to illustrate the contamination problem.)

All of the experiments described above were performed inside either a standard laboratory chemical fume hood or an enclosure specially designed for handling nanoparticles. Most scientists would assume that the handling of nanopowders inside such hoods would offer them good protection from exposure; our research indicates that such hoods certainly offer a good deal of protection, but that airborne ENPs can escape at measurable levels. This points to the need for a very careful application of engineering controls if they are to be effective at controlling nanopowder exposures, and for the evaluation of their performance. The performance of such fume hoods is discussed in detail in Chapter 9, and techniques for evaluating their performance are discussed in Chapter 8.

6.4.2   Liquid Suspensions

The handling of ENPs in liquid suspension is inherently preferable to handling them in dry powder form, since the potential for exposure is much lower. Liquid suspensions, however, still present opportunities for exposure, depending on how they are handled and processed. Circumstances where exposure can occur include *spraying*, *agitation*, and *accidental spills*.

6.4.2.1 Spraying   This is the classical method to aerosolize liquids. Figure 6.4, taken by the US Works Progress Administration in the early 1940s, shows a worker spray-painting parts (note the primitive respirator and local exhaust system). Here we are primarily concerned with subsequent exposures to the ENPs suspended in the liquid, but exposures to the liquid itself and its evaporated vapors may also be of

FIGURE 6.4   Worker spray painting parts, c. 1940. Photo used by permission Wikimedia/ U.S. Works Progress Administration.

TABLE 6.1 Drying times for water droplets
at 50% relative humidity and 293 K

Droplet diameter (μm)	Drying time (s)
0.01	1.6×10^{-6}
0.04	1.4×10^{-5}
0.1	4.7×10^{-5}
0.4	3.6×10^{-4}
1.0	1.7×10^{-3}
4.0	0.024
10	0.15
40	2.3

Adapted from Hinds (1999).

concern; this is discussed below. In industrial operations, spraying is typically used when it is desired to coat a surface with a controlled amount of the coating material. The droplet diameters produced by typical industrial spraying operations can vary from submicrometer to thousands of micrometers, depending on the nozzle design and operating conditions. When ENP suspensions are sprayed, the liquid may be water or any number of other liquids, such as solvents with higher vapor pressures than water to speed the drying time.

Once spraying occurs, we must be concerned with the potential for exposure. Exposure is complicated by the fact that as soon as it is created, the liquid aerosol starts evaporating; thus a worker in the vicinity may be exposed to the liquid aerosol containing the ENPs, the ENPs in dry form if evaporation is complete, the liquid vapors, or some combination of all three.

The evaporation rate of a liquid is affected primarily by the liquid's vapor pressure, which in turn is affected by the ambient temperature and the concentration of the vapor in the surrounding air. If the droplet's liquid is water, the surrounding vapor concentration is typically indicated by the relative humidity. In addition, for droplets, the curved surface of the liquid allows molecules to escape more easily than if the surface was flat; this is known as the Kelvin effect. The result is that droplets tend to evaporate very quickly, with smaller droplets evaporating more quickly than larger ones. Table 6.1 lists the drying times for water droplets of various diameters at 50% relative humidity (Hinds, 1999).

The effect of inherent liquid vapor pressure on drying time is illustrated in Table 6.2 (Hinds, 1999). Solvents with higher vapor pressures than water (such as ethyl alcohol in the table) evaporate even faster than water, while liquids such as oils with very low vapor pressure (such as DEHP in the table) dry very slowly.

We can now return to the question of exposure. As indicated in Tables 6.1 and 6.2, workers in the immediate vicinity of a spraying operation may be exposed to dry ENPs or partially-evaporated droplets containing ENPs. From a health standpoint, the primary effect of the variability in evaporation state is the size of the particle/droplet when it is inhaled; this will determine the deposition pattern of the particle in the respiratory system, as discussed in Section 4.2. In addition, the evaporation process

TABLE 6.2   Drying times (s) for different liquids in vapor-free air at 293 K

Droplet diameter (μm)	Ethyl alcohol	Water	Mercury	Di-ethyl hexyl phalate (DEHP)
0.01	4×10^{-7}	2×10^{-6}	0.005	1.8
0.1	9×10^{-6}	3×10^{-5}	0.3	740
1	3×10^{-4}	0.001	1.4	3×10^{4}
10	0.03	0.08	1200	2×10^{6}
40	0.4	1.3	2×10^{4}	4×10^{7}

Adapted from Hinds (1999).

complicates the measurement of exposure, since the particle size distribution is changing over time. As discussed in Section 7.2, instruments that measure real-time exposure have response times varying from one second to several minutes; instruments with longer response times may be measuring the diameters of droplets that are actually changing during the measurement period, and thus not represent the actual exposure of a worker who may be exposed to the spray aerosol within a few seconds after it is formed. In addition, the carrier liquid may itself be toxic, and the worker may be exposed to it in both the liquid and evaporated vapor forms; although beyond the scope of this book, this must be addressed by the occupational hygienist.

6.4.2.2 Agitation   In addition to spraying, liquid suspensions of ENPs can be subject to several other forms of agitation, which may or may not release ENPs to the air. A very common form of agitation is simple stirring, which is done to ensure even mixing of materials throughout a liquid or to enhance a chemical reaction. In a research facility, such stirring is typically done in an open container such as a beaker or flask, but in manufacturing plants, stirring more commonly occurs in closed vessels. If stirring occurs in an open vessel, and the stirring is energetic enough to create "waves" on the liquid surface, then liquid aerosols may be produced; if these conditions are not met, little or no ENP release will occur.

Another form of agitation common to research laboratories is sonication, which uses ultrasonic energy to break up particle agglomerates that tend to form when nanoparticles remain suspended in a liquid for some period of time. Probe sonicators (Fig. 6.5), in particular, may release a significant number of droplets containing ENPs (Taurozzi et al., 2010).

6.4.2.3 Other Operations Involving Liquids   Besides spraying, most industrial operations involving liquids are done in closed systems or in open vessels, where the liquid may evaporate but the ENPs will remain suspended. In these cases, the greatest risk for exposure is from incidental or accidental skin contact with the liquid, or from an accident where the liquid is spilled onto a surface. In the case of a spill, contact can occur during the spill cleanup, or the liquid may evaporate, leaving the ENPs behind where they can be resuspended and inhaled or come into skin contact. Exposures such as these are best handled by the strict use of good work practices at the facility; this topic is discussed in detail in Section 9.3.

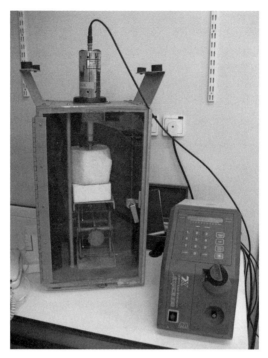

FIGURE 6.5   Probe sonicator. Photo used by permission Wikimedia/Eyal Bairey.

6.4.3   ENPs Bound to a Solid

ENPs can be solidly bound to a solid primarily in two ways, that is, mixed into the bulk of the solid, as in a composite consisting of ENPs and a polymer, or in a surface coating which is applied and hardens. In either case, once the nanoparticles are bound in a solid form, the potential for exposure is greatly reduced.

For materials in this category, the greatest potential for exposure occurs during the manufacture of the ENP-containing material. For example, the manufacture of a nanoparticle-polymer composite typically involves the feeding of the two materials in dry form into an extruder, where the polymer is melted under high temperature and pressure and mixed with the nanomaterial. The feeding of the dry nanopowder to the extruder can be a very dusty process, with great potential for particle release. This scenario is described in great detail in the extruder case study presented in Section 8.10.

Once the material is made, nanoparticle release can occur when energy is applied to it, for example, by sanding, cutting, drilling, or simple weathering. Bello and colleagues (2009) measured NP release during the wet and dry cutting of CNT and carbon fiber-containing advanced composites and found a significant particle release ($>10^5$ particles/cm^3) for dry cutting but not wet cutting. Only a few free fibers (1–4 fibers/cm^3) were detected in the air samples. Methner et al. (2012) measured particle release during the wet sawing and surface grinding of polymer-carbon nanofiber composites; they also found that wet sawing did not release appreciable numbers of airborne NPs, but the surface grinding created airborne concentrations as high 500,000 particles/cm^3. Although the number of free fibers was not quantified, they were detected in TEM images.

In addition, nanoparticles may be released when a composite or coated material is disposed of at its end of life. As discussed in Section 6.5.2, little is known at this time about this potential exposure path.

6.5 ENVIRONMENTAL EXPOSURES TO ENGINEERED NANOPARTICLES

As used here, "environmental exposures" refers to two scenarios: (1) at a manufacturing site, releases of ENPs from the facility into the air, water, or solid waste and (2) any exposures occurring in the life cycle of an ENP-containing product after that product leaves the manufacturing site. The second category includes consumer use of the product and its end-of-life processing.

6.5.1 Environmental Releases

At this time, little is known about actual releases of ENPs from manufacturing facilities. This is largely due to the lack of specific emission regulations in the United States, the European Union, or other countries where manufacturing is now occurring. For this reason, we can only include a fairly general discussion of the factors that are likely to be important.

6.5.1.1 Air Emissions Few regulations now exist that specifically limit the air emissions of ENPs. One exception is Japan; the following statement is included in a notification issued by the Director of Labour Standards Bureau, Ministry of Health, Labour and Welfare (2008):

> In principle, manufacturing devices shall have a sealed structure. If it is difficult, install a local exhaust system. The outlet of a local exhaust system to outside must be attached with a high performance filter, and installed local exhaust systems shall be maintained and inspected periodically.

In addition, some guidance documents recommend that control technologies be implemented to reduce emissions to the air. For example, Texas A&M University (2005), in their *Interim Guideline for Working Safely with Nanotechnology*, under "strategies to control exposure to nanoparticles" lists "local exhaust ventilation, with HEPA filtration." Although no details are given as to how HEPA filtration specifically is to be used for this purpose, they do proceed with a fairly comprehensive discussion of HEPA design and performance. HEPA filter design and use are fully discussed in Chapter 10.

However, since most available guidelines are meant to help protect workers from exposure, they typically discuss the types of exposure controls covered in Chapter 9 and do not mention environmental exposure controls. As a common example, research laboratories typically rely on chemical fume hoods to protect their workers (sometimes a poor practice when it comes to ENPs, as discussed in Chapter 9), but the exhaust from those hoods is very rarely filtered before being released to the environment.

6.5.1.2 Water Emissions   As of this writing, there are very few published studies concerning the emission of ENPs into wastewater and their subsequent transport. Schwegmann and Frimmel (2010) reviewed the various pathways by which ENPs can enter the water cycle. They include emission into wastewater as by-products of manufacturing, removal of textile coatings, cosmetics, sunscreens, and so on, during washing, runoff from outdoor applications such as spraying, painting, and so on, weathering of NP-containing structures, and transfer from air to soil and water by direct deposition or rain.

Once ENPs enter the groundwater, their fate must be studied, which includes their transport through the system and their potential effects on microorganisms in the soil and micro- and larger organisms in water. The concerns are summarized by Fent (2010):

> To date, little is known about the occurrence, fate and potential effects of ENP in the environment. The potential for effects on ecosystems will depend not only on the amount of nanoparticles emitted but also on their physico-chemical characteristics (size, surface/volume ratio, shape, chemical composition), which can be influenced by environmental parameters … The exposure of organisms will largely depend not only on the concentration, size and surface characteristics of dissolved NP but also on adsorption and aggregation phenomena … Adsorption to environmental media such as sediments, biofilms and microsurface layers of open waters may influence the bioavailability and thus exposure. To date, studies devoted to evaluate the bioavailability in a systematic fashion are lacking.

6.5.2   Exposures Through a Product's Life Cycle

Thus far, this chapter has focused on possible exposures arising out of the manufacturing process. In the broader field of environmental health, however, the impact of toxic materials throughout their lifetime is receiving considerable attention; this is the purview of life cycle assessment (LCA). There is a growing recognition that the impact of new ENPs must be assessed throughout their life cycle; the term used for this analysis is LCA. The International Standards Organization (ISO, 2006) defines life cycle as "consecutive and interlinked stages of a product system, from raw material acquisition or generation from natural resources to final disposal," and LCA as "compilation and evaluation of the inputs, outputs and the potential environmental impacts of a product system throughout its life cycle."

In addition to manufacturing, the primary life cycle stages of a product are consumer use and end-of-life disposal. The most likely routes of exposure for these two stages are briefly discussed here.

6.5.2.1 Consumer Use Hansen et al. (2008) suggest that the likelihood of exposure to nanoparticle-containing consumer products can be predicted by "…dividing all nanomaterials into three overall categories: (1) in the bulk, (2) on the surface, and (3) as particles." Examples of each category include nanocomposites (bulk), coatings containing NPs (surface), and NPs suspended in liquids (particles). At the time their article was published, the Woodrow Wilson Project on Emerging Technologies (WWICS, 2009) had 580 products listed in their inventory of nanotechnology-enabled consumer products. Of the products that could be categorized, the vast majority contained NPs in particle form, with 13% suspended in solids, 19% bound to surfaces, and 37% suspended in liquids. Only about 1% were powders containing free NPs, and about the same number were in bulk form.

Following the above categorization, Hansen et al. further categorize nanoparticle-containing consumer products by likelihood of exposure, again with three categories:

1. *Expected to cause exposure.* Products that either "…require direct human exposure or make it possible." The major categories meeting this definition are nanoparticles suspended in liquids and airborne nanoparticles.
2. *May cause exposure.* Products where the nanoparticles are not meant to be released, but ordinary wear and tear when in use may lead to a release. The primary category here is nanoparticle-containing surface coatings.
3. *No expected exposure to the consumer.* Products where the nanoparticles are encapsulated into a solid product.

Although the process outlined above seems to make sense and results in a *qualitative* estimate of the likelihood of exposure, the next step, actually *quantifying* any such exposure, is much more difficult. Three possible approaches are to actually measure exposures while the product is being used (using techniques similar to those used in quantifying occupational exposures), measure exposures in the laboratory under simulated use conditions, or use models to predict exposure. Hansen attempted to calculate likely exposures for four different scenarios using available models for skin absorption, airborne concentration, and so on, but the results obtained are likely to have significant uncertainties, since the models used make many assumptions concerning "typical" conditions of product use.

There are no studies in the published literature using actual exposures, but several simulations have been conducted. For example, Chen et al. (2010) conducted tests of a spray can containing TiO_2 nanoparticles and used for bathroom disinfection "…in a laboratory room with 9-ft ceilings that was remodeled to simulate a home environment." They found that the liquid in the spray evaporated rapidly, as expected, and that the resulting airborne TiO_2 concentration was approximately 10^5 particles/cm^3, with a count median diameter of 75 nm and a geometric standard deviation of 2.3. They conclude that "the results suggest that consumers could be exposed to a significant concentration of airborne TiO_2 particles while using TiO_2 in a spraying application."

Nazarenko and colleagues evaluated exposures from the use of both nanoparticle-containing spray products (Nazarenko et al., 2011) and cosmetic powders

(Nazarenko et al., 2012). In the spray products study, five products from the Woodrow Wilson inventory and a similar non-nanoparticle product were tested by spraying them toward a mannequin head placed in a biological safety cabinet. In the cosmetic powders study, three pairs of nanoparticle and non-nanoparticle products were applied to a female mannequin head inside a glove box.

The airborne particles from all of the spray products, nanoparticle-containing and not, showed a wide size distribution over the entire nanoparticle size range, with total concentrations in the 10^3 particles/cm^3 range. The authors comment that "a very interesting and rather surprising result of the study is the detection of nano-sized (1–100 nm) particles not only in nanotechnology-based, but also in regular (non-nanotechnology) products…" They conclude:

> On the basis of the obtained data, it is difficult to conclude whether the nanoparticles released during the product use are actually engineered nanoparticles that were incorporated into the product or they are derivatives from natural product ingredients, such as from herbal oil emulsification, or they are particles from product carrier liquid.

As with the spray products, the airborne particles released from the application of cosmetic powders all showed a wide range of particle diameters across the entire 10 nm to 10 μm size range, with no significant differences between the nanometer- and non-nanometer-sized powders. Electron microscopy confirmed that the powders were in a highly agglomerated state, which would account for the similarity in size distributions between the two product categories. They conclude that "our findings on potential nanomaterial inhalation exposure due to the use of actual consumer products emphasize that properties and effects of the pure nanomaterial ingredients cannot be used to predict actual consumer exposures and resulting health effects."

6.5.2.2 *End of Life Disposal*   To the best of our knowledge, there are no published studies that evaluate actual exposures during the disposal of nanoparticle-containing products. The potential for particle release likely would depend on the same categorization scheme as described above for consumer use, but exposure pathways are likely to be very complex in various disposal environments, such as landfills (Shatkin and Davis, 2008). Much work must be done before any conclusions can be reached regarding population exposures during this life cycle phase.

REFERENCES

Bello D, Wardle BL, Yamamoto M, Guzman de Villoria R, Garcia EJ, Hart AJ, Ahn K, Ellenbecker MJ, Hallock M. Exposure to nanoscale particles and fibers during machining of hybrid advanced composites containing carbon nanotubes. J Nanopart Res 2009;11:231–249.

Burgess WA, Ellenbecker MJ, Treitman RD. *Ventilation for Control of the Work Environment.* New York: Wiley-Interscience; 2004.

Chen BT, Afshari A, Stone S, Jackson M, Schwegler-Berry D, Frazer DG, Castranova V, Thomas TA. Nanoparticles-containing spray can aerosol: characterization, exposure assessment, and generator design. Inhal Toxicol 2010;22:1072–1082.

Fent K. Ecotoxicology of engineered nanoparticles. In: Frimmel FH, Niessner R, editors. *Nanoparticles in the Water Cycle—Properties, Analysis and Environmental Relevance*. Berlin: Springer-Verlag; 2010. p 184.

Hansen SF, Michelson ES, Kamper A, Borling P, Stuer-Lauridsen F, Baun A. Categorization framework to aid exposure assessment of nanomaterials in consumer products. Ecotox 2008;17:438–447.

Hinds WC. *Aerosol technology—properties, behavior, and measurement of airborne particles*. New York: Wiley-Interscience; 1999.

International Standards Organization [ISO]. Environmental management—life cycle assessment—principles and framework. Geneva: International Organization for Standardization; 2006. ISO 14040:2006(E).

Japanese Ministry of Health, Labour and Welfare. Notification on present preventive measures for the prevention of exposure at workplaces manufacturing and handling nanomaterials. Tokyo: Japanese Ministry of Health, Labour and Welfare; 2008. Notification No. 0207004.

Maynard A, Baron P, Foley M, Shvedova A, Kisin E, Castranova V. Exposure to carbon nanotube material: aerosol release during the handling of unrefined single walled carbon nanotube material. J Toxicol Environ Health A 2004;67:87–107.

Methner M. Engineering case reports: effectiveness of local exhaust ventilation (LEV) in controlling engineered nanomaterial emissions during reactor cleanout operations. J Occup Environ Hyg 2008;5:D63–D69.

Methner M, Crawford C, Geraci C. Evaluation of the potential airborne release of carbon nanofibers during the preparation, grinding, and cutting of epoxy-based nanocomposite material. J Occup Environ Hyg 2012;9:308–318.

Nazarenko Y, Han TW, Lioy PJ, Mainelis G. Potential for exposure to engineered nanoparticles from nanotechnology-based consumer spray products. J Expo Sci Environ Epidemiol 2011;21:515–528.

Nazarenko Y, Zhen H, Han T, Lioy PJ, Mainelis G. Potential for inhalation exposure to engineered nanoparticles from nanotechnology-based cosmetic powders. Environ Health Perspect 2012;120:885–892.

Schwegmann H, Frimmel FH. Nanoparticles: interaction with microorganisms. In: Frimmel FH, Niessner R, editors. *Nanoparticles in the Water Cycle—Properties, Analysis and Environmental Relevance*. Berlin: Springer-Verlag; 2010. p 165.

Shatkin JA, Davis JM. Alternative approaches for life cycle risk assessment for nanotechnology and comprehensive environmental assessment. In: Shatkin JA, editor. *Nanotechnology—Health and Environmental Risks*. Boca Raton (FL): CRC Press; 2008. p 113.

Taurozzi JS, Hackley VA, Wiesner MR. *CEINT/NIST Protocol for Preparation of Nanoparticle Dispersions from Powdered Material Using Ultrasonic Disruption*. Gaithersburg (MD): Center for the Environmental Impact of Nanopmaterials, Duke University, and the U.S. National Institute for Standards and Technology; 2010.

Texas A&M University. *Interim Guideline for Working Safely with Nanotechnology*. College Station (TX): Texas A&M University; 2005.

Tsai SJ, Ashter A, Ada E, Mead J, Barry C, Ellenbecker MJ. Airborne nanoparticle release associated with the compounding of nanocomposites using nanoalumina as fillers. Aerosol Air Qual Res 2008a;8:160–177.

Tsai SJ, Ashter A, Ada E, Mead J, Barry C, Ellenbecker MJ. Control of airborne nanoparticle release during compounding of polymer nanocomposites. Nano 2008b;3:1–9.

Tsai SJ, Hofmann M, Hallock M, Ada E, Kong J, Ellenbecker MJ. Characterization and evaluation of nanoparticle release during the synthesis of single-walled and multi-walled carbon nanotubes by chemical vapor deposition. Environ Sci Technol 2009a;43:6017–6023.

Tsai SJ, Ada E, Isaacs J, Ellenbecker MJ. Airborne nanoparticle exposures associated with the manual handling of nanoalumina and nanosilver in fume hoods. J Nanopart Res 2009b;11:147–161.

Woodrow Wilson International Center for Scholars [WWICS]. 2009. The project on emerging technologies. WWICS. Available at www.nanotechproject.org/. Accessed November 22, 2009.

7

EVALUATION OF EXPOSURES TO ENGINEERED NANOPARTICLES

Current knowledge on the toxicity of engineered nanoparticles was discussed in Chapter 5. The fact that a substance has the *potential* to cause damage does not necessarily mean, however, that actual damage will occur. This requires that a person be *exposed* to the toxic material, and that the material enters the body as a result of that exposure and travels to a body organ where damage can occur. Thus, as discussed in Chapter 4, *exposure characterization* and ultimately *exposure assessment* must receive attention comparable to that given toxicology.

7.1 CURRENT KNOWLEDGE CONCERNING EXPOSURE TO ENGINEERED NANOPARTICLES

Given the current level of interest in the topic, the published peer-reviewed literature concerning exposure to engineered nanoparticles (ENPs)—either for workers or the general public—has many fewer publications than those studying biological effects. Several groups across various countries are performing such exposure evaluations in various settings; however, the results are still limited to a few specific exposure scenarios. This chapter will introduce the current knowledge concerning methodologies being used to characterize exposure to ENPs; the use of these methodologies will be discussed in some detail in Chapter 8.

Exposure Assessment and Safety Considerations for Working with Engineered Nanoparticles, First Edition. Michael J. Ellenbecker and Candace Su-Jung Tsai.

Given the likely toxicity of carbon nanotubes (CNTs), there probably have been more workplace exposure papers published for this nanomaterial than any other; a few examples will be given here. Maynard et al. (2004) reported on the results of a NIOSH study that found almost no release of airborne fibers when CNTs were removed from a high-pressure carbon monoxide (HIPCO) reactor and transferred into a secondary container. The highest mass concentration measured during the transfer procedure was low ($53 \mu g/m^3$) when compared to allowable exposures for other forms of carbon such as graphite (permissible exposure limit (PEL) of $5 mg/m^3$). A NIOSH technical report (Baron et al., 2003) found similar results at several other locations manufacturing small-walled CNTs (SWCNTs) by the laser ablation and HIPCO methods. Tsai et al. (2009a) evaluated particle releases from a furnace producing carbon nanotubes by chemical vapor deposition and found large numbers of both fibrous and nonfibrous carbon particles in the furnace exhaust.

Han et al. (2008) conducted air monitoring for multiwalled CNTs (MWCNTs) during several procedures in a research laboratory. The MWCNTs had undergone several postproduction processing steps, including dispersion and functionalization with sonication, before monitoring. They found that levels of MWCNTs during a blending procedure generated relatively pure airborne MWCNTs at concentrations between 172.9 and 193.6 particles/cm³ using NIOSH Method 7402 to count the fibers under transmission electron microscope (TEM). The addition of an enclosure and exhaust ventilation reduced levels between 0.018 and 0.05 particles/cm³.

A NIOSH Nanotechnology Research Center evaluation (Methner, 2008) at a facility producing metal alloy spheres approximately 15–50 nm in diameter found concentrations as high as 17,000 particles/cm³ during reactor cleanout, with no exhaust ventilation. The use of local exhaust ventilation (LEV) located adjacent to the cleanout operation reduced the maximum concentration to only about 2000 particles/cm³.

In 2007, NIOSH conducted their first ENP health hazard evaluation at a university research lab producing polymer composites using carbon nanofibers (CNFs) (Methner et al., 2007). According to the report, "the following specific processes were identified for further evaluation:

1. Chopping of extruded composite material (proprietary formula) containing CNFs.
2. Transferring CNFs (approximately 1 lb) from a plastic receptacle outside a laboratory hood to a small beaker for weighing inside the hood.
3. Transferring and mechanically mixing CNFs with acetone inside a 5-gal mixing vessel positioned on the floor outside the hood without LEV.
4. Cutting composite material using a water-cooled, dust-suppressed table saw (wet saw).
5. Manually sifting oven-dried, epoxy-coated CNFs on an open benchtop to remove large clumps."

Emission controls on this process were found to be good in general. Two activities, that is, weighing and mixing of the CNFs prior to processing, and the wet-saw cutting of the nanocomposite product, produced airborne nanoparticle concentrations that were elevated slightly above background levels.

7.2 EXPOSURE TO ENGINEERED NANOPARTICLES BY INHALATION

At this time, there is not a consensus on the best way to evaluate airborne exposures to ENPs. Almost all existing toxic aerosol exposure standards are based on mass concentration; the notable exception is asbestos, where exposure standards are based on number concentration. Occupational hygienists prefer mass-based aerosol sampling because the sampling and sample analysis are straightforward; you pre-weigh a filter, pass the air sample through the filter at a calibrated air flow, postweigh the filter, and calculate the measured concentration in, for example, milligrams per cubic meter. Let us take a closer look at mass sampling for ENPs.

7.2.1 Mass Sampling

There is a consensus among most experts that mass sampling may *not* be appropriate for ENPs, for the simple reason that by definition is such particles have very little mass. An example of this was given in Chapter 3, where in our thought experiment of cutting a 1-m block of graphite into 1-nm cubes illustrated the very low mass concentrations that would result. As another example, let us consider SWCNTs. Some material safety data sheets for CNTs suggest that the PEL for graphite or carbon black should apply, since CNTs are after all just carbon. As discussed in Chapter 3, the PEL for respirable graphite in air is $5\,mg/m^3$; if we assume a typical size for a CNT, we can calculate the number concentration corresponding to $5\,mg/m^3$.

A typical SWCNT is a cylinder with a diameter of approximately 1–3 nm (depending on the manufacturing method) (Lu and Liu, 2006) and a length that is highly variable. For the purpose of this exercise, let's assume that the tube is fairly long—1 mm (one million times the diameter). The density of a SWCNT depends on its diameter; for our chosen diameter, it is approximately $2.5\,g/cm^3$ (Laurent et al., 2010). We can now calculate the volume and mass of this cylinder. The volume V of a cylinder of diameter d and length L is:

$$V = \frac{\pi d^2 L}{4} \tag{7.1}$$

Converting our dimensions to meters and substituting them into Equation (7.1) yields:

$$V = \frac{\pi (10^{-9}\,m)^2 (10^{-3}\,m)}{4} = 7.8 \times 10^{-22}\,m^3$$

The mass m of a cylinder equals its volume times its density, ρ:

$$m = \rho V \tag{7.2}$$

Substituting the values into Equation (7.2) yields the mass of a single SWCNT:

$$m = (2500\ kg/m^3)(7.8 \times 10^{-22}\,m^3) = 2.0 \times 10^{-18}\ kg = 2.0 \times 10^{-12}\,mg$$

We can now calculate the number of SWCNTs per cubic meter, n, at the assumed mass concentration, c, of $5\,mg/m^3$:

$$n = \frac{c}{m} \qquad (7.3)$$

Again substituting the assumed values:

$$n = \frac{5 \ mg\,/\,m^3}{2.0 \times 10^{-12} \ mg\,/\,SWCNT} = 2.6 \times 10^{12} \ SWCNT\,/\,m^3 = 2.6 \times 10^6 \ SWCNT\,/\,cm^3$$

This is not as astronomically high as the number concentration for $1\,nm$ graphite cubes calculated in Chapter 3, since our example CNT is much larger, but it still is an astonishingly large number concentration, more than seven orders of magnitude higher than the current US asbestos PEL of $0.1\,fibers/cm^3$ (OSHA, 2006). The animal toxicology data discussed in Chapter 5 suggest that MWCNTs may behave in a similar fashion to asbestos; if this is even remotely true, the above calculation suggests that an allowable airborne mass concentration for CNTs should be about seven orders of magnitude lower than $5\,mg/m^3$, or about $0.5\,ng/m^3$!

After reviewing the above calculations, one might come to the conclusion that, indeed, the graphite PEL is not appropriate for CNTs, but we could use the lower value calculated above and still be able to rely on mass sampling. The problem with this approach is that typical background dust concentrations in the air are on the order of $50\,\mu g/m^3$, or $50{,}000\,ng/m^3$, so that trying to detect the mass of CNTs would be akin to searching for the proverbial needle in a haystack.

Some have argued that you can surmount this problem by collecting a filter sample and analyzing it, not for total dust, but for elemental carbon, which presumably would largely be tied up in CNTs. The problem with this approach is the limit of detection (LOD) for elemental carbon analytical methods. NIOSH Analytical Method 5040 (NIOSH, 2003) for elemental carbon gives an LOD of approximately $2\,\mu g/m^3$ for a sample volume of $960\,l$ (a sampling rate of $2\,L/min$ for a full 8-h day); a repeat of the aforementioned calculations yields an equivalent maximum number concentration LOD of $1000\,SWCNT/cm^3$, still four orders of magnitude higher than the asbestos PEL.

Despite the high number levels implicit in a mass-based CNT standard, in November 2010, NIOSH issued a draft Current Intelligence Bulletin (CIB), "Occupational Exposure to Carbon Nanotubes and Nanofibers," (NIOSH, 2010) that proposed an 8-h time-weighted average recommended exposure limit (REL) for CNTs of $7\,\mu g/m^3$ based on elemental carbon as measured by the NIOSH Analytical Method 5040. The final CIB, published in April 2013 (NIOSH, 2013) states "previously in a 2010 draft of this CIB for public comment, NIOSH indicated that the risks could occur with exposures less than $1\,\mu g/m^3$ but that the analytic limit of quantification was $7\,\mu g/m^3$. Based on subsequent improvements in sampling and analytic methods, NIOSH is now recommending an exposure limit at the current analytical limit of quantification of $1\,\mu g/m^3$." This lowers the equivalent number concentration to $500\,SWCNT/cm^3$ in our example explained earlier, 5000 times the asbestos PEL. In addition, this REL assumes a sample flow of $4\,L/min$ for

8 h, rather than the more standard 2 L/min—a value that many sampling pumps cannot deliver. This proposed REL is discussed in full detail in Section 11.1.

Although the earlier discussion focused on CNTs, the same argument holds to a great extent for any particle in the nanometer size range. Obviously, the lower the toxicity of a given ENP, the higher the allowable exposure, and the more feasible mass sampling might be. One example of this approach is the NIOSH REL for nanometer-sized titanium dioxide (TiO_2), published in 2011 (NIOSH, 2011). In this document, NIOSH defines "ultrafine" TiO_2 as "the fraction of respirable particles with a primary particle diameter of <0.1 µm (100 nm)." For ultrafine TiO_2, the NIOSH REL is 0.3 mg/m^3 and the REL for fine TiO_2 ($d_p > 100$ nm) is 2.4 mg/m^3, both as time-weighted averages for up to 10 h/day and a 40-h week. The publication documents in some detail the rationale for these RELs, also discussed in Section 11.1, but the point here is that the 0.3 mg/m^3 value is well within the capabilities of standard occupational hygiene gravimetric sampling methods, which is not the case for the CNT REL.

If mass sampling is not feasible for ENPs, the other possibilities are total number of particles in a certain size range, the particle size distribution, their surface area in a particular size range, and the surface area distribution. Much of the ENP toxicity literature seems to correlate adverse effect with particle surface area, so a surface area metric may make sense as an indicator of exposure. In addition, particle number concentration may make the most sense for nanoparticles where the actual number of particles that get deposited in the respiratory system is the best indicator of likely toxic effect. This is true, for example, for any particle such as CNTs that appear to cause fibrosis, since in this case, each particle depositing in the alveolar region disrupts alveolar macrophages, causing the fibrotic response. Likewise, the number of very small particles picked up by the olfactory nerve is probably more important than their actual size or surface area. In 2005–2007, Wittmaack and the Oberdörsters had an interesting interplay concerning the relative importance of surface area and number (Oberdörster et al., 2005, 2007b; Wittmaack, 2007). We will return to this topic in Section 11.1, where we discuss occupational exposure limits for ENMs.

7.2.2 Surface Area Measurement

If surface area plays a strong role in nanoparticle toxicity, then it may seem to make sense to measure particle surface area concentration and to develop exposure indices based on surface area. In response to this perceived need, instrument manufacturers have begun to market surface area monitors. Both the TSI Model 3550 (Fig. 7.1a) and TSI Aerotrak 9000 (Fig. 7.1b) surface area monitors do not measure the total surface area of the airborne nanoparticles; rather, they use the ICRP lung deposition model discussed in Chapter 4 to measure the lung-deposited nanoparticle surface area either in the tracheobronchial or the alveolar region. The major difference between the instruments is that the 3550 is a stationary instrument (must be plugged into an outlet) while the Aerotrak 9000 is powered by battery and thus more portable. The instruments work on the same principle, that is, diffusion charging followed by electrometer detection, and measure the deposited surface area for particles with diameters from 10 to 1000 nm; they have a response time as low as 1 s.

(a)   (b)

FIGURE 7.1 TSI Nanoparticle Surface Area Monitors: (a) Nanoparticle Surface Area Monitor (NSAM) Model 3550; (b) AEROTRAK™ 9000 Nanoparticle Aerosol Monitor. Images courtesy of TSI, Inc.

While in principle surface area may be important based on toxicity, the measurement of total surface area in the manner used by the instruments described earlier may be misleading for polydisperse aerosols. The reason is that, since particle surface area varies by the square of particle diameter, any measure of total surface area will be dominated by the larger particles, whereas the toxicity of the smallest particles with the largest surface area to volume ratio may in fact be more important. Since these instruments sample particles with diameters up to 1000 nm, this problem is likely to be significant in some circumstances even taking into account the deposition fraction as predicted by the ICRP model.

7.2.3 Number Concentration Measurement

Before describing the instruments now available for determining the number concentration of aerosol particles in the nanometer size range, a brief history of aerosol instrument development will be presented. This will give the reader an appreciation of how we got to where we are today.

7.2.3.1 Impaction In the earliest days of occupational hygiene, back near the start of the twentieth century, efficient sampling filters were not available, and occupational hygienists relied almost exclusively on measuring number concentration to characterize aerosols. Three widely used types of instruments were developed early in the 1900s—the konimeter, the Owens jet dust counter, and the Greenburg-Smith impinger (Marple, 2004). All three instruments were variations of inertial impactors, where airflow streamlines are turned by one mechanism or another and the particles with sufficient inertia cross streamlines and impact on a collection surface. These early impaction devices could only collect particles in the micrometer range, but more recent developments, described below, have adapted the impaction principle to the collection of nanoparticles.

The workhorse instrument was the impinger. Impingers collect particles by drawing air through a jet suspended in water; the particles impact on the interface

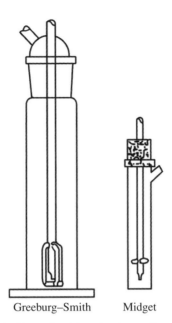

Greeburg–Smith   Midget

FIGURE 7.2   Greenburg-Smith and midget impinger. Drawing courtesy of the U.S. Environmental Protection Agency.

between the air bubble and the water and are collected for later counting using an optical microscope. As the original version, the Greenburg-Smith impinger, was large, bulky, and very difficult to use, the midget impinger, with all dimensions reduced by a factor of ten, was developed in 1932 (Hatch et al., 1932) (Fig. 7.2). Measurements reported in the literature using devices of this type typically report concentrations in millions of particles per cubic foot (mmpcf). Although in theory the collected particles larger than about 0.3 μm in diameter could be sized under the microscope to attain a size distribution, this was rarely reported in the early literature.

The use of these early instruments spurred research on improved devices, leading to the first cascade impactors, invented by May in the 1940s. May first developed a four-stage impactor (Fig. 7.3), where the particles were deposited on glass slides (May, 1945). By weighing the slides before and after use, the mass distribution could be determined. In order to determine the particle sizes collected on each stage, May calibrated his instrument by sizing particles under the microscope—a task that was apparently looked on with disfavor. A quote from May, as reported by Marple (2004), is instructive on this point (and also on the status of women in science in the 1940s):

> To return to the original calibration of the impactor, this had to be done by extensive counting and sizing of sampled droplets under the microscope. This is tedious and needs mental application but I find no sympathy with the horrified revulsion with which such work is regarded by so many. May I suggest that laziness or perhaps the prospect of a Ph.D. has inspired much lengthy and highly expensive research to automate the rapid

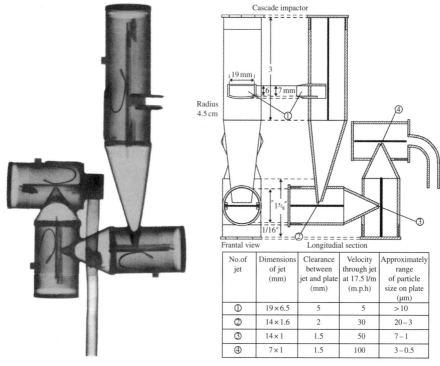

No.of jet	Dimensions of jet (mm)	Clearance between jet and plate (mm)	Velocity through jet at 17.5 l/m (m.p.h)	Approximately range of particle size on plate (μm)
①	19 × 6.5	5	5	> 10
②	14 × 1.6	2	30	20 – 3
③	14 × 1	1.5	50	7 – 1
④	7 × 1	1.5	100	3 – 0.5

FIGURE 7.3   Original May cascade impactor. Reprinted with permission from May (1945). Copyright 1945 IOP Publishing Ltd.

evaluation of a particle deposit (with dubious accuracy?). It could well be that 100 man-hours of research have been spent to save one person-hour at the microscope. Why spend $50,000 on a pile of black boxes which often go wrong when for the same money you can get 10 years out of a prettier and replaceable assistant who can also make coffee?

May further refined his design, resulting finally in the development of his "ultimate" impactor in 1975 (Fig. 7.4a), which still used microscope slides for the impaction surfaces but now increased the number of stages to 8 (May, 1975).

A major limitation in the use of traditional impactors is the range of particle sizes that can be evaluated. The impaction efficiencies of the May ultimate impactor stages are shown in Figure 7.4b, where the fraction of particles collected is plotted against particle aerodynamic diameter. The impactor designer attempts to make this curve as steep as possible, in order to divide the incoming aerosol into two distinct size fractions; the particle size collected with 50% efficiency is called the *cut diameter*, d_{50}:

$$d_{50}\sqrt{C_c} = \left[\frac{9\pi\eta D_j^3 \left(\mathrm{Stk}_{50} \right)}{4\rho_p Q} \right]^{1/2} \tag{7.4}$$

(a)

(b)

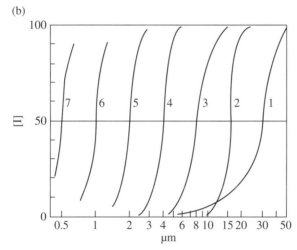

FIGURE 7.4 May "ultimate" impactor: (a) photo of impactor, (b) collection efficiency curves. Reprinted with permission from May (1975). Copyright 1975 Elsevier, Inc.

where, η = air viscosity (1.81×10^{-5} Pa·s at STP), D_j = impactor jet diameter (m), Stk_{50} = Stokes number for 50% impaction efficiency (a property of the impactor design), ρ_p = particle density (kg/m^3), Q = impactor air flow (m^3/s), and C_c = Cunningham correction factor (used to correct Stokes' law for small particles).

In a cascade impactor, since all of the other variables are constant, the jet diameter is made smaller in succeeding stages in order to decrease the cut diameter. This technique can only be pushed so far; however, the smallest particle size that can be collected is limited to those particles that have sufficient inertial properties, usually an aerodynamic diameter of about 200–300 nm. Another approach for a given impactor design would be to increase the air flow, since this will decrease the cut diameter for each stage; this also has practical limitations, since only so much air can be drawn through a given device. The cut diameters for a typical cascade impactor, made by Anderson (Fig. 7.5), are

FIGURE 7.5 Anderson cascade impactor. Used with permission, the PUrE Consortium, http://www.pureintrawise.org/?page=research&theme=particulates.

TABLE 7.1 Aerodynamic cut diameters for the Anderson impactor at 28.3 and 60 L/min

	Aerodynamic cut diameter (μm)	
Impactor stage	$Q = 28.3$ (L/min) or (1 ft³/min)	$Q = 56.6$ (L/min) or (2 ft³/min)
0	9.0	6.4
1	5.8	4.1
2	4.7	3.3
3	3.3	2.3
4	2.1	1.5
5	1.1	0.78
6	0.70	0.49
7	0.40	0.28

Cut diameter is the diameter collected by the impactor stage with 50% efficiency.

shown in Table 7.1. As the table indicates, under normal flow (28.3 L/min, or 1 ft³/min), the smallest cut diameter is 400 nm, and doubling the flow reduces this to 280 nm.

The earlier discussion illustrates the inability of a typical cascade impactor to determine particle size distributions in the nanometer size range. There are at least two approaches around this problem, however. The first is to use very small jet diameters; this is the approach used in the Moudi impactor. The second approach is to operate the impactor at low static pressures. This has the effect of decreasing the gas viscosity and this, according to Equation (7.4), will reduce the cut diameter. This is the approach taken by the Dekati impactor.

The MSP Corp. Model 125B Nano-Moudi-II™ micro-orifice impactor is the latest version of this instrument, developed at the University of Minnesota (Chow et al., 2008). It operates at 10 L/min and has 13 impactor stages giving cutoff diameters from 10 to 10,000 nm. A unique feature of this device is that the stages rotate during use, so that the particle deposition is spread out across the impactor surface, which helps maintain a steep particle collection curve. A disadvantage of this device is that the collected particles are measured gravimetrically, so that a sample must be collected for a long enough time to obtain measurable mass on all of the stages, and daunting task for the smaller particles; another disadvantage is that the micro-orifices are prone to clogging.

The Dekati electrical low pressure impactor, ELPI+™, combines the principle of low-pressure impaction with the use of electrometers for the automatic sizing of particles from 6 to 10,000 nm in 14 size channels (Chow et al., 2008). Upon entering the device, the particles are charged to a known level and passed through the low-pressure impactor; the stages are electrically neutral, so the particles deposit based on their aerodynamic diameter, but the electrometers on each stage record the particle deposition by detecting their charge. The current on each stage is proportional to particle number concentration and size, so that the signals from each stage can be combined into a particle size distribution. The instrument is portable but somewhat heavy, weighing 22 kg (48 lb) with its inlet preseparator impactor.

7.2.3.2 Filtration

Research on filter design led to the development in the 1950s of air sampling filters that could efficiently collect aerosol particles of all sizes. Efficient filters made it relatively easy and straightforward to measure mass concentration, simply by pre- and postweighing the filters. Consequently, for the past 50 years or so, filters have become the workhorse industrial hygiene tool for evaluating worker exposure to aerosols, and most exposure limits for aerosols are given in terms of allowable mass concentration. Filters are used for gravimetric analysis, as described earlier. In addition, filters can be used to collect particles for subsequent analysis by light or electron microscopy; this application is described in the following section.

7.2.3.3 Light Scattering

Advances in the understanding of the physics of light scattering in the 1950s (Sinclair, 1950; van de Hulst, 1957) led to the development by the 1970s of commercially available instruments to automatically count and size aerosol particles (Waggoner and Charlson, 1976; Willeke and Liu, 1976). Such instruments are widely available today, but their primary limitation from a nanoparticle perspective is that the smallest particle measurable is limited by the wavelength of visible light (400–700 nm). Particles much smaller than this size are not reliably sized; in addition, the maximum particle size measured is a function of the aerodynamic properties of the instrument's sampling inlet.

Most commercial instruments typically size and count particles from about 0.5 to 20 μm; a common example of such an instrument is the Grimm Model 1.109 Portable Aerosol Spectrometer (Fig. 7.6) (Tiwary and Colls, 2004). In instruments of this type, the particles are passed through a sensing volume, which is illuminated by a

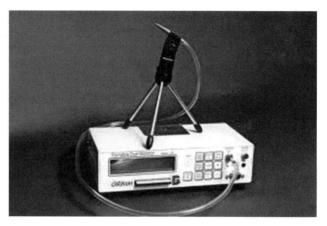

FIGURE 7.6 Grimm optical aerosol spectrometer and dust monitor. Photo courtesy of Grimm Aerosol Technik GmbH & Co. KG.

light source; earlier instruments used incandescent lamps, but newer instruments rely on laser light. The light scattered at a particular angle is collected and focused on a photodetector, which produces an electrical current proportional to the number of photons collected. The number concentration is determined by the pulse rate, and the particle size is determined by the pulse height. The primary limitation of this type of instrument is that the amount of light scattered by a particle is a function not only of its physical size but also of several other properties, including shape and index of refraction. Another limitation is the maximum concentration for which only one particle at a time will be in the sensing volume; if two particles are sensed simultaneously, they will be erroneously recorded as one larger particle—this is termed coincidence error (Hinds, 1999). The Grimm instrument has typical specifications for this type of instrument; it sizes particles from 0.25 to 32 μm in 31 size channels, with a maximum total particle concentration of 2000 particles/cm³.

Because of the lower size limit where light scattering is effective, this technique cannot be used directly for counting and sizing nanoparticles. However, the condensation particle counter (CPC) uses light scattering by first growing the particles to the micrometer size range by condensation; this technique is described in the following sections.

7.2.3.4 Particle Acceleration A relatively new class of particle sizing instruments relies on the time of flight of aerosol particles in an accelerating flow field. Smaller particles will rapidly adjust to the new air velocity, while larger particles will take longer to accelerate. A widely used instrument based on this principle is the TSI Aerodynamic Particle Sizer (APS) (Fig. 7.7); a similar instrument, the Aerosizer, is manufactured by Amherst Process Instruments, Inc. In these instruments, the aerosol-laden air is passed through an acceleration nozzle. At the nozzle exit, the aerosol particles pass by two laser beams, and particles are detected by collecting scattered light in two photometers; the time delay between the two particle detections is then

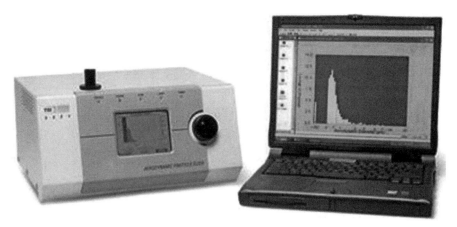

FIGURE 7.7   Aerodynamic Particle Sizer® (APS™) Model 3321. Photo Courtesy of TSI Inc.

used to calculate the particle's aerodynamic diameter. The advantage of this approach over simple light scattering is that the laser photometers are used just to detect each particle, not to size it.

Unfortunately, the smallest particle size that can be detected using acceleration is about 500 nm, since particles smaller than this have insufficient inertia and are accelerated to the higher flow field velocity in a very short time. Frequently, however, it is desirable to collect particle size data across both the nanometer and micrometer size range, to get a complete aerosol particle size distribution. In that case, the APS can be used in conjunction with other instruments, such as the scanning mobility particle sizer (SMPS) and fast mobility particle sizer (FMPS) described later; this approach is routinely followed in our laboratory, as discussed in Chapter 8.

7.2.3.5 Currently   Available   Instruments   for   Counting   and   Sizing Nanoparticles   There are a growing number of commercial instruments available for counting and sizing particles in the nanometer size range. They rely on one or a combination of several techniques, as described in the following sections.

7.2.3.6 Condensation Particle Counters   As discussed earlier, light-scattering devices are not effective in the nanometer size range. A CPC surmounts this difficulty by first growing nanometer-sized particles to the micrometer range and then counting them by light scattering. CPCs were first developed by John Aitken (1888); the earliest models were called condensation *nuclei* counters, since they were designed to count the nanometer-sized atmospheric particles that serve as condensation nuclei for raindrops. An early model, the Pollack CNC (Pollak and Metneiks, 1960), is shown in Figure 7.8, and a schematic of a typical CPC used for nanoparticle measurements, the TSI Model 3007, is shown in Figure 7.9.

A CPC utilizes supersaturation and condensation to grow nanoparticles to a size that can be counted optically. As shown in Figure 7.9, the aerosol is first drawn into

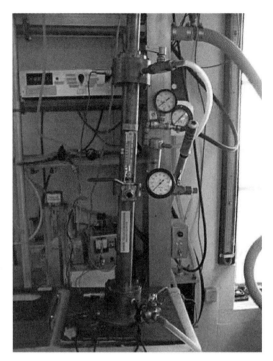

FIGURE 7.8   Pollack condensation particle counter. Photo courtesy of U.S. Department of Transportation.

a chamber that is heated and lined with a porous medium that is saturated with water, alcohol, or another easily evaporated liquid. The air becomes saturated with the liquid vapor in this chamber, and is then drawn into a condenser where the temperature is reduced, causing the vapor pressure to become super-saturated. The excess vapor condenses on the particles in the air stream, causing them to grow. These larger particles are then counted by light scattering.

A modern CPC such as the Grimm 5.403, the Kanomax 3800, or the TSI 3009 is hand-held, battery-operated, and measures total particle concentrations as high as 10^7 particles/cm^3 in the nanoparticle size range in a short time period—typically 1 s. Since a CPC does not give particle size information, it is most useful as a survey instrument, where the user walks through the industrial process of interest, looking for sources of elevated nanoparticle release as indicated by the CPC. These nanoparticle sources can then be investigated in more detail using the more sophisticated instruments described in the following text.

7.2.3.7   Scanning Mobility Particle Sizers   If an aerosol is processed to remove all but a narrow range of particle sizes and then passed through a CPC, the particle concentration in that narrow size range can be counted. This is the principle of operation of the SMPS. Commercially available SMPS instruments perform the size discrimination step using a differential mobility analyzer (DMA), shown in

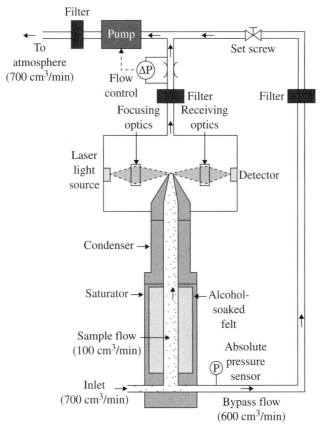

FIGURE 7.9 TSI condensation particle counter (CPC) Model 3007 schematic. Drawing courtesy of TSI Inc.

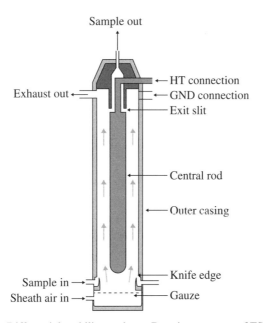

FIGURE 7.10 Differential mobility analyzer. Drawing courtesy of TSI Incorporated.

Figure 7.10. In a typical DMA, a polydisperse aerosol in the nanometer size range is drawn into the instrument and passed through a bipolar charger, which imparts charges onto the particles based on the Boltzmann equilibrium charge distribution. In the nanometer size range, most particles will have one positive charge, one negative charge, or no charge.

Once charged, the particles enter the main chamber, consisting of a grounded housing and a center rod maintained at a high negative voltage. The negatively charged particles are repelled from the rod and move toward the housing at their terminal electrical mobility velocity, which is a function of their size. Only particles within a narrow range of electrical mobility, and thus a narrow diameter range, will travel the exact distance to reach the exit slot. This narrow size range is then passed to the CPC, where they are grown and counted. By varying the voltage, different size particles can be selected for counting. Thus, this class of instruments is called a "scanning" mobility particle sizer because a complete size distribution is determined by scanning across the size range of interest. A typical size distribution thus takes a certain time period, typically one to several minutes depending on the number of size channels measured.

Versions of the SMPS are marketed by several companies, including TSI, Kanomax, and Grimm. Commercial models vary in the overall particle size range detected, the number of size channels, the maximum measurable particle number concentration, the type of CPC employed (long vs. short column, water vs. alcohol, etc.), and the physical size of the device. TSI markets several SMPS models, ranging from the hand-held, battery-operated Nanoscan SMPS (Fig. 7.11a) to the widely used "workhorse" Model 3034 (Fig. 7.11b) and the multicomponent "top of the line" Model 3936 (Fig. 7.11c). The performance characteristics of these three models, along with the Grimm Model SMPS+C and Kanomax Portable Aerosol Measuring Spectrometer (PAMS), are compared in Table 7.2. Also included in the table is the TSI FMPS, described later.

The Nanoscan and the PAMS are battery-operated, portable SMPS devices. The Nanoscan measures NPs from 10 to 350 nm in 13 size channels, while the PAMS has two operating modes—one that measures NPs from 14.5 to 863 nm in 14 size channels and one that measures NPs from 10 to 433 nm in 27 size channels. The Grimm SMPS+C takes its portable CPC and adds a DMA to provide particle sizing capability in a portable instrument. It measures nanoparticles from 5 to 1100 nm in up to 255 size categories. It is somewhat heavier than the Nanoscan and Kanomax (13 kg vs. 9 and 4.5 kg), which is somewhat compensated for by its wider size range and larger number of size channels.

The TSI Model 3034 is also "portable" but only if it is placed on a cart; its dimensions are 44 cm wide × 35 cm deep × 59 cm high (17 × 14 × 23 in.), it weighs 28 kg (61 lb), and it must be plugged into a 110 V outlet. This instrument might best be described as "transportable," and many researchers have taken it into the field. The Model 3936 is a component system, as seen in Figure 7.11c; the electrostatic classifier with its DMA column is on the left and the CPC is on the right. This is definitely a laboratory instrument, used primarily in sophisticated aerosol physics laboratories.

(a)   (b)

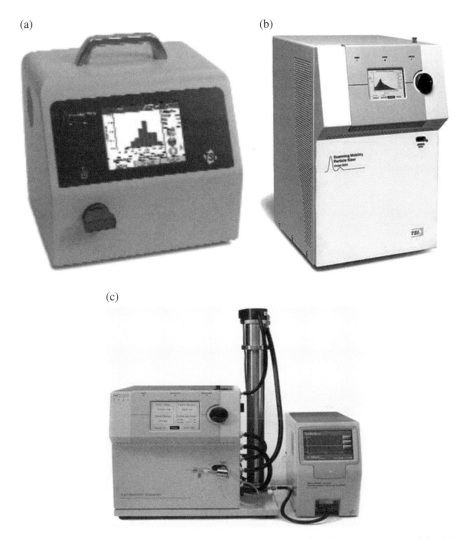

(c)

FIGURE 7.11   TSI scanning mobility particle sizer™ SMPS™ Spectrometers models. (a) Nanoscan SMPS Nanoparticle Sizer Model 3910; (b) SMPS spectrometer Model 3034; (c) SMPS spectrometer Model 3936. Photos courtesy of TSI Incorporated.

7.2.3.8   Fast Mobility Particle Sizer   The TSI Model 3091 FMPS spectrometer is similar in size and shape to the Model 3034 SMPS but operates on a different principle. It is based on research performed at the University of Tartu in Estonia in the early 1990s.[1] It is similar in operation to a DMA in that particles are passed through negative and positive corona chargers to produce a charge distribution that is a function of the particle size. The positively charged particles are then introduced into

[1] TSI FMPS product brochure: http://www.tsi.com/uploadedFiles/Product_Information/Literature/Spec_ Sheets/3091FMPS.pdf

TABLE 7.2 Characteristics of commercial nanoparticle spectrometers

Manufacturer/ model	Particle size range (nm)	Number of size channels	Sample collection time	Form factor; use	Weight, kg (lb)
Grimm/SMPS+C	5–1100	255		Portable; field	13 (29)
Kanomax PAMS	14.5–863	14	"minutes"	Portable; field	4.5 (10)
	10–433	27	"minutes"		
TSI/3910 Nanoscan	10–350	13	1 min	Portable; field	9 (19)
TSI/3034	10–487	54	3 min	Transportable; field	28 (61)
TSI/3936	2.5–1000	Up to 167	16 s	Components laboratory only	Varies
TSI/3091 FMPS	5.6–560	32	1 s	Transportable; field	32 (70)

an annular space surrounding a positive electrode and move under the electrostatic force toward 22 grounded electrometers. The electrostatic force is countered by the Stokes drag force, so the particles move at a constant velocity toward the electrometers. Particles with a high electrical mobility are collected on the upper electrometers, while those with lower mobilities penetrate to deeper ones. When a particle strikes an electrometer, it discharges and the current levels detected on all of the electrometers are processed to yield 32 size classifications over two orders of magnitude, from 5.6 to 560 nm. One important advantage of the FMPS over all SMPS models is the sample time, which is only 1 s; this allows the measurement and characterization of rapidly changing nanoparticle concentrations. A second advantage is that liquids are not needed to operate the instrument.

From the outside, the FMPS looks like the Model 3034 SMPS. Its width and depth are the same as the Model 3034, but it is taller (70 cm, 28 in.) and heavier (32 kg, 70 lb). The maximum detectable concentration ranges from 10^7 particles/cm^3 for the smallest particles to 10^5 particles/cm^3 for the largest particles, and the minimum detectable concentration ranges from about 100 particles/cm^3 for the smallest particles to about 1 particle/cm^3 for the largest particles. This instrument was adopted for use by many research groups a few years after the SMPS; for example, it has been a mainstay of the research performed by the authors. Typically, researchers collect samples with multiple instruments to provide a broad range of particle concentration and size distribution such as using the FMPS in parallel with the TSI APS (Fig. 7.7); both instruments can be operated with a sample time of 1 s, and their combined size range is 5.6 nm to 20 μm. Figure 7.12 shows a typical sampling arrangement the authors use for these instruments; the instruments are carried on a cart, and the APS and the data collection laptop are placed on top of the FMPS. Detailed examples on the use of the FMPS for exposure characterization are presented in Chapter 8.[2]

[2] The detailed examples in this book typically use FMPS data, because that is the "workhorse" instrument used in our laboratory. The examples are meant to illustrate the type of data that can be collected by any nanoparticle spectrometer, and are not meant to promote this particular instrument.

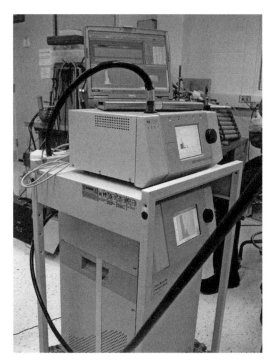

FIGURE 7.12   A sampling cart with the FMPS, APS, and data collection laptop. Photo by S. Tsai.

Given the variety of instruments described above, the question arises as to the comparability of results obtained when using them. This was investigated by Asbach and colleagues (2009), who performed side-by-side comparison testing of the FMPS, the MMPS+C, and two different TSI SMPS models incorporating different CPCs. The instruments were challenged with both nanometer-sized sodium chloride (NaCl) particles and diesel exhaust soot particles, which fall in the nanometer size range. In general, all three of the SMPS instruments measured similar concentrations and size distributions, but the FMPS size distribution was consistently shifted toward smaller particle sizes. In addition, the SMPS instruments responded consistently for the two tested aerosols; the FMPS responded differently, however, with narrower distributions and lower concentrations for NaCl compared to the SMPSs and the opposite result for diesel soot. Asbach et al. suggest "one possible reason may be the different morphologies of NaCl and diesel soot particles, causing differences in particle charging." Jeong and Evans (2009) found that the SMPS consistently measured lower particle concentrations than the FMPS, which they attributed to "diffusion losses during transport through the sampling system of the SMPS...."

Grimm recently developed an instrument based on similar electrostatic principles as the FMPS, the SMPS+E. As of this writing, this instrument has not been evaluated in any peer-reviewed publications.

7.2.4   Conversion between Number, Surface Area, and Mass Concentrations

There are several reasons why it would be advantageous to be able to convert between the different ENP concentration metrics. For example, various published reports concerning exposure measurements to the same type of NP may use different exposures metrics, making it difficult to directly compare the results. If further work indicates that surface area concentration is the most biologically relevant metric for some particular NP, it will be necessary to convert both mass and number concentrations to surface area.

Most toxicology studies report animal inhalation doses on a mass basis, again making it difficult to compare the animal doses to human exposures based on count data. On the other hand, some recent toxicology studies have reported dose in terms of number concentration—an example is a recent carbon black NP study (Kim et al., 2011); since most if not all previous studies of this material have reported mass doses (see, e.g., Vesterdal et al., 2010), direct comparison with this new study is not possible. One recent study (Gangwal et al., 2011) converted number-based human exposure data to mass for comparison to *in vitro* testing.

For a single spherical particle, conversion between diameter, surface area, and volume is simply geometry; if the particle density is known, the particle mass can likewise be calculated. Conversions are more complicated when the individual particles are not spherical and when a range of particle sizes is present. Let us first consider nonspherical particles. Many NPs are actually agglomerates of smaller, spherical particles, as shown in Figure 2.5. Even assigning a single "diameter" to such an agglomerate is problematic. An automated instrument such as an SMPS may count and size a sample of such particles, but conversion to surface area, volume, or mass is problematic. The first issue is the "diameter" of the particle, as registered by the SMPS, that is, the size channel it is assigned by the instrument. This relates to the measurement method used by the instrument. The SMPS charges each particle by passing it through a corona and then it transports the particle through a DMA to a CPC. Only particles with a certain narrow range of electrical mobility pass through the DMA, so the measured diameter is actually the "electrical mobility diameter," that is, the diameter of a spherical particle with the same electrical mobility as the measured particle. For an agglomerate such as that in Figure 2.5, this is primarily a function of the total charge acquired by the particle and its mass. Since an agglomerate has more surface area than an equivalent spherical particle, it likely will collect more charges (Wang et al., 2010), thus having a higher electrical mobility and be detected and placed in a smaller size channel. For any automated instrument, it is likely that the sizing of an agglomerate will have some errors; since surface area is a function of the square of the diameter and volume/mass a function of the cube, any uncertainties in diameter estimation will be magnified when converted to these other quantities.

The ease in converting from number concentration to surface area or mass concentration for a polydisperse aerosol depends on the information available. Many aerosols exhibit a log-normal size distribution; which simplifies the conversion since

Gaussian statistics can be used. If a size distribution by count is log-normal, a count median diameter (CMD) and geometric standard deviation can be calculated; the Hatch-Choate conversion equations can then be used to calculate the diameter of average surface area or average mass, which can then be multiplied by the total number concentration to give the surface area or mass concentration. Readers interested in pursuing this approach should consult a standard aerosol text, such as Hinds (1999). An alternative approach is to make the conversion on individual size channels; this approach is illustrated by Tsai (2013); an even simpler approach is to just assume that a polydisperse aerosol can be characterized by a single diameter and use that diameter for the conversion (Gangwal et al., 2011).

7.2.5 Particle Characterization

All of the instruments described in the earlier text attempt to measure the size of nanoparticles, using one property or another of that particle. None of them measure the actual physical dimensions of the particles, and the measured particle sizes are reported as a diameter (or, more commonly, a range of diameters) as if each particle is a perfect sphere. We know, however, that most ENPs most definitely are not spherical; in fact some of them, such as carbon nanotubes, are far from spherical. In addition, none of the instruments described above identify the materials that comprise the particles; this can be very important, for example, in distinguishing particles released from a particular source from the preexisting background (see Chapter 8).

It is thus very important in many circumstances to have the ability to determine particle shape, or *morphology*, and *elemental composition*. For the past 100 years, particles in the micrometer size range have been examined under light microscope, in order to determine their size and shape. Micrometer-sized particles, however, are too small to be detected by light microscopy; here, we must rely on the electron microscope. Both the scanning electron microscope (SEM) and TEM can be used to analyze nanoparticles—the choice depends largely on instrument availability.

When using an SEM, the particles are collected from the desired location onto a sampling filter, using standard air sampling methodologies. Typically, the battery-operated air sampling pump used in occupational hygiene work is used, but larger sampling systems can be used if area samples are collected. Since an SEM has a very narrow depth of field, the nanoparticles must be collected on a very flat surface; the recommended filter for this is the capillary pore membrane filter (Hinds, 1999)—a common brand is the Nucleopore filter, although they are also manufactured by other companies. In order to ensure high collection of the smallest nanoparticles, the smallest available pore size, currently 200 nm, should be used; the disadvantage of using such small pores is that the pressure drop across the filter is very high, which puts a strain on the sampling pump.

Once the sample is collected, a portion of the filter is coated with a conductive metal, usually gold, and placed in the SEM for analysis. The sample preparation and microscopic analysis require considerable skill and must be performed by a person with training and experience in this methodology. One big advantage of using SEM is that it produces a three-dimensional image of the collected particles, as seen in Figure 7.13 which shows a typical result collected from our laboratory.

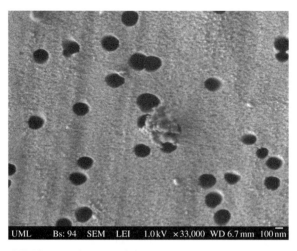

FIGURE 7.13   SEM image of nanoparticles collected on a capillary pore membrane filter. Image by S. Tsai.

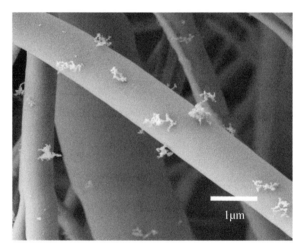

FIGURE 7.14   SEM image showing fibers in a fabric filter sample along with collected nanoparticles. Image courtesy E. Ada, University of Massachusetts Lowell.

As mentioned in the earlier texts, when the objective is to just examine nanoparticles it is preferable to use a capillary pore membrane filter, but SEM can also be used to investigate three-dimensional structures incorporating ENPs. For example, the authors have investigated the use of fabric filters to collect ENPs, as discussed in Section 10.6. Figure 7.14 is an SEM image of a close-up of fibers in a fabric sample and the collected nanoparticles; images such as this assist us in understanding the particle collection mechanisms at work in the filter.

TEM is a very useful technique for evaluating particle morphology, for elemental analysis, and for particle counting and sizing. Since TEM relies on analyzing high-energy electrons that pass *through* the sample, a filter or other solid substrate cannot be used to collect the particles; instead, a TEM grid is used to collect the particles for analysis. The primary difficulty with using TEM for analysis is getting the nanoparticles onto the grid. Two broad approaches can be used, that is, indirect methods and direct methods. In the past, indirect methods have usually been used, as for example in NIOSH Method 7402, "Asbestos by TEM" (NIOSH, 1994). Indirect methods, however, have several serious disadvantages that include the following:

- Sample preparation is difficult and labor-intensive. The samples are collected on a filter, and the filter must be processed to remove the particles and resuspend them onto the TEM grid. Typically, particles are removed from the filter with a liquid, the liquid is evaporated, and the concentrated particles are transferred to a grid. As an example of the complexity involved, NIOSH Method 7402 lists 20 different items of required equipment, ranging from tweezers and Petri dishes to a vacuum evaporator.
- The sample preparation process may affect the particle size distribution, particularly with respect to agglomerates, which are very prevalent in nanoparticle samples. Existing agglomerates may be broken apart by the particle removal process.
- On the other hand, the sample concentration process, where the liquid used to collect the particles from the filter is evaporated and the particles are transferred to the grid, may create new agglomerates.

The shortcomings of the indirect methods have led scientists to experiment with various methods to directly deposit nanoparticles from the air onto a grid. We have developed the system shown in Figure 7.15, which uses a standard occupational hygiene personal sampling pump and filter cassette. A TEM grid is taped to the surface of a capillary pore membrane filter, the pump is operated, and nanoparticles are drawn past the grid in the air flow and migrate to the grid by diffusion.

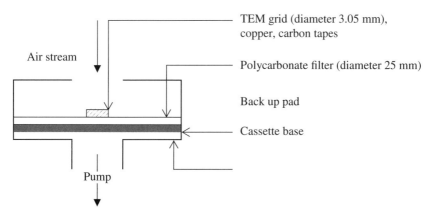

FIGURE 7.15   Nanoparticle sampler.

The equations used to predict collection by diffusion are presented in the Appendix 7.A.1 and typical sampling conditions are discussed in Chapter 8. Diffusion is also used by the Naneum wide range aerosol sampler (Gorbunov et al., 2009), but in this device, the NPs are deposited onto wire meshes rather than TEM grids. There are no published studies on the collection efficiency of a diffusive sampler, although we have described its use in several of our papers (Tsai et al., 2008, 2009b, 2012).

Two other techniques have been adapted to commercially available instruments. The first is thermal precipitation, where a temperature gradient is used to drive nanoparticles to the surface of the grid (Gonzalez et al., 2005). The thermophoretic sampler, manufactured by the Fraunhofer Institute for Toxicology and Experimental Medicine (ITEM), operates on this principle; in our experience this instrument is fairly bulky and difficult to use, but future versions may improve on its performance. More recently, RJ Lee, working with Colorado State University (Thayer et al., 2011), has developed a prototype of a thermal sampler that is about the size of a matchbox, battery-operated, and able to be worn by a worker. As of this writing, no plans have been finalized for the commercialization of this device.

NIOSH (Miller et al., 2010) and a group of Swiss researchers (Fierz et al., 2007) have performed research on the use of electrostatic precipitation to move nanoparticles to a grid. NIOSH has licensed their invention to Dash Connector Technology, Inc., whose first commercially available device, the ESP Nano, is on the market. This device is hand-held, battery-operated, and very easy to use. Grimm has developed a similar device, which is designed to be used with their DMA, and InTox Products has marketed an ESP TEM device for several years. One potential problem with this approach is that it relies on the charging of the particles in order to collect them in an electric field; since it is very difficult to induce a charge on the very smallest nanoparticles, the collected sample may underestimate the lower end of a sample size distribution. Our diffusion device, on the other hand, favors small particles at the expense of larger ones, which have lower diffusion coefficients; the same should be true of the thermal precipitators.

As of this writing, there has been no published comparison of the ability of these various techniques to collect a representative nanoparticle sample onto a TEM grid. They all show promise and offer distinct advantages over the indirect methods.

7.3 DERMAL EXPOSURES TO ENGINEERED NANOPARTICLES

As discussed in Section 4.3, the skin as a route of exposure has historically received much less attention than inhalation. The same is true of exposure assessment techniques. The literature has thousands of articles dedicated to air monitoring, airborne exposure limits, and so on, but very little can be found concerning dermal exposure measurement. For example, Volume 2 of the latest Patty's Industrial Hygiene, titled *Evaluation and Control* (Rose and Cohrssen, 2011a), is 755 pages long, and exactly ten pages are devoted to dermal exposures. Volume 1, *Hazard Recognition* (Rose and Cohrssen, 2011b), devotes less than a page to skin sampling methods.

With regard to nanoparticles, the literature to date has focused on dermal penetration and toxicity (see Section 4.3) and has not considered exposure measurement. Needless to say, no standard reference methods for evaluating nanoparticle dermal exposure have been adapted or even proposed. As a recent report from the Organization for Economic Cooperation and Development (OECD, 2009) stated:

> Sampling and analytical methods developed for assessing dermal exposures to chemicals have not been evaluated for their applicability to characterize dermal exposures to nanomaterials in the workplace.

Nonetheless, we can review the available methods for evaluating dermal exposures, since they likely will have applicability to nanoparticles. Schneider et al. (2000) reviewed the available measurement methods; the most useful ones are as follows:

- *In situ* measurement using fluorescence
- Wet wiping of the skin
- Particle stripping with adhesive tape
- Skin washing
- Direct collection on a patch placed on the skin
- Vacuuming the skin (e.g., with a personal sampling pump, closed face filter cassette, and tubing).

All of these methods have deficiencies, and none of them has been used to evaluate nanoparticle dermal exposure, as reported in the published literature. We have had some success in using fluorescent-tagged NPs, and detecting them on gloves and clothing using ultraviolet light, as shown in Figure 7.16. We have found this technique to be useful for documenting that skin or clothing contamination occurs, but we have not used it to quantify skin exposure. It is also useful as a training technique, as discussed in Section 9.3.

(a)   (b)

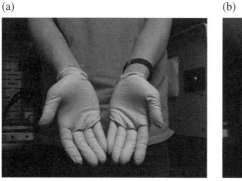 

FIGURE 7.16 Fluorescent nanoparticle glove contamination: (a) normal light, no particles visible; (b) same gloves with NPs visible under ultraviolet light. Photos by S. Tsai.

Concerns regarding surface contamination from settled dust have led to the development of "wipe test" methods, where a measured surface area is dry-wiped with a filter or fabric and the collected particles analyzed using standard techniques (Rose and Cohrssen, 2011b). These techniques have become popular for detecting possible asbestos surface contamination, and a standard methodology (ASTM D6480–05(2010)) has been developed; in this case, the presence of asbestos fibers is measured using TEM.

Wipe test methods were developed for flat, bare surfaces. It is unclear how effective a wipe test would be in removing nanoparticles from a damp, hairy irregular skin surface. The methods selected by Schneider et al. as listed earlier are likely to be more successful.

7.4   EVALUATION OF EXPOSURES IN AQUATIC ENVIRONMENTS

7.4.1   Introduction

At this time in the evolution of ENP health and safety research, almost all of the attention has focused on worker exposures—primarily by inhalation but also by dermal contact. As more nanotechnology-enabled products are brought to market, however, we can expect the focus to move toward exposures to the general public and environmental impacts of those products. In anticipation of this, we have included here a short overview of instruments and methods now available to evaluate ENPs in aquatic environments.

Hassellov and Kaegi (2009) have written an excellent overview of methods available for analyzing ENPs in aquatic environments, and this section primarily summarizes the material present in that chapter. As with airborne nanoparticles, those in aquatic environments can be classified as natural, industrial, or engineered nanoparticles. There are many types of natural NPs in water, including metal oxides and humic materials. Industrial NPs are the same as those found in air; particles such as from diesel exhaust and power plant combustion, are typically released into the air and transferred to water systems by diffusion and rainout. ENPs may also be released into the air and transferred to water systems, or they may be suspended in industrial waste streams.

ENPs suspended in water are a *colloid* system, defined as a stable suspension of particles in water, and there is a great deal of scientific literature concerning colloids. Oberdörster et al. (2007a) reviewed the literature concerning the environmental fate and transport of colloids and concluded:

> The techniques, models and knowledge currently available in relation to colloid behaviour, are all applicable to engineered nanoparticles and will provide a useful knowledge base on which to investigate such environmental issues.

7.4.2   Sample Collection

All of the particle characteristics discussed in Chapter 3 (size distribution, morphology, etc.) are just as relevant to NPs in water as to NPs in air and, consequently, a range of instruments is available to measure them. One major difference between air and water is the sample preparation stage. As discussed earlier, real-time instruments for air analysis always sample directly from the air, but for water such *in situ* measurements are rare. According to Hassellov and Kaegi (2009),

> Environmental colloidal systems are dynamic and consist of continuous particle formation and dissolution, removal and adsorption of organic matter and microbial degradation. It is well known that natural NPs in the colloidal state are easily disturbed in water, groundwater, soil and sediment during sampling, handling and storage. Changes can include agglomeration, microbial growth, degradation or adhesion to containers. It is, therefore, desirable to use in situ techniques that remove the sampling step, but there are almost none available that can provide more than very basic information. Instead, efforts have been put into sampling of the colloids as gently as possible, minimizing handling and storage to avoid perturbations. These procedures optimized for natural NPs should be suitable also for manufactured NPs in the same matrices.

7.4.3   Measurement Methods

7.4.3.1   Light Scattering   There are more light scattering techniques available for particles in water than in air. Classical light scattering is used and typically has the same particle size limits as for particles in air. CytoViva, Inc. has developed specialized darkfield microscopy techniques that allow the measurement of Rayleigh scattering from nanoparticles in liquid samples down to the nanometer size range. They also use hyperspectral imaging to identify specific types of nanoparticles.[3]

In addition, two techniques are available that rely on the Brownian motion of small particles in water. *Dynamic light scattering* (DLS) relies on short-term changes in scattered light intensity due to changes in interference patterns between adjacent particles as they move by Brownian motion. A major limitation of DLS is that, in the nanometer size range, the amount of light scattered by a particle varies with the sixth power of diameter. This means that in a polydisperse aerosol, the signal from larger particles easily masks the signal from smaller particles, thus biasing the measured size distribution toward the larger particles.

A newer technique, *nanoparticle tracking analysis* (NTA), tracks the motion of a collection of particles by illuminating them with a laser, viewing them with a light microscope, and taking a high-speed movie of the particles (Fig. 7.17). The positions of the particles in adjacent movie frames are analyzed, and their root mean square motion is used to calculate their diffusion coefficient (see Appendix 7.A.1). NanoSight (Amesbury, UK) makes several versions of an NTA instrument, all of which use an integrated camera that captures video at 30 frames/s and automatically measures particle size distributions over a range of diameters approximately from 10 to 1000 nm.[4]

7.4.3.2   Fractionation and Separation   Several techniques can be used to separate suspended particles into different size fractions, which can then be quantified by a second technique such as light scattering. *Filtration* and *centrifugation* are widely

---

[3] http://www.cytoviva.com/
[4] http://www.nanosight.com/

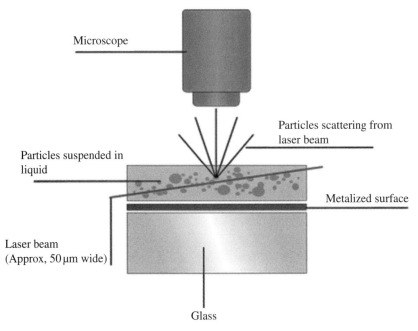

FIGURE 7.17   Schematic of nanoparticle tracking analysis system. Reproduced courtesy of Malvern Instruments.

used, but skill is required in order to avoid changing the particle properties (e.g., by agglomeration). *Field-flow fractionation* (FFF) is a collection of techniques that apply a force field to the particles and measure their response. The most common forms are to use hydrodynamic fluid flow (flow FFF) or centrifugal force (sedimentation FFF), where the field is applied to the fluid containing the NPs and the particles' response distributes them across the field as a function of their diameter. *Chromatography*, using porous or capillary media, can separate particles since different diameters move through the medium at different speeds. Finally, *electrophoresis* can be used to fractionate charged particles in an electric field (similar to the DMA for NPs in air). Zeta potential, a measure of the potential difference between the surface of a particle and the liquid in a colloid and a measure of the colloid's stability, can be measured using instruments based on electrophoresis.

7.4.3.3 Particle Morphology and Elemental Analysis   Electron microscopy can be used to characterize ENPs in water, just as in air. The critical difference lies in the sample preparation. For SEM analysis, NPs can be collected onto the surface of a capillary pore membrane filter, just as in air analysis; the water filtration process must be done with care, however, to avoid the formation of agglomerates, pore clogging, overloading of the filter surface, and so on (Hassellov and Kaegi, 2009). For TEM analysis, an indirect method must be used to transfer NPs from the aqueous sample to the grid; according to Hassellov and

Kaegi (2009), "…for TEM investigations, a proper sample preparation procedure is of the highest importance, and often determines the quality of the analysis obtained afterwards."

7.4.4   Exposure Characterization in Aquatic Environments

Exposure assessment in aquatic environments is complicated by several factors not present in the air environment, including:

- There are many target species in water, while in air we are worried primarily about human exposures.
- The medium—water—can be highly variable (pH, chemicals in solution, suspended organic matter, etc.), while the atmosphere is much more stable.
- Water can have significant effects on the NPs themselves. In particular, NPs, with their very high surface-to-volume ratio, can dissolve, which means that particle concentrations and size distributions will change over time.
- NPs can move back and forth between the water and the subsurface soil or sediment.

Hassellov and Kaegi recommend a three-tiered approach to exposure assessment in the complex aquatic environment: (1) initial material characterization, (2) fate and behavior assessment in a range of relevant environmental media conditions, and (3) detailed exposure characterization in the media conditions of interest. They summarize the difficulties as follows:

> To accommodate all analytical requirements, strategies need to be developed and methods optimized for characterization of NPs (free, deposited, aggregated) and dissolution products in a range of complex matrixes (soil, sediment, sewage sludge, and biological tissue). Hence, it is not only advanced analytical instrumentation that is required but, as importantly, sampling, handling, storage protocols, extraction and digestion methods also need to be optimized and validated.

It seems fair to conclude that, at this time, instrumentation and exposure assessment methods for ENPs in air are more advanced than for ENPs in water. As the focus of environmental health professionals expands from workers to the environment, we can expect this imbalance to begin to be corrected.

APPENDIX 7.A.1:   Derivation of Nanoparticle Displacement by Diffusion

When collecting nanoparticles through Brownian motion, air velocity is critical to maximize particle collection, since the lower the air velocity, the higher the net mean displacement by Brownian motion. The interaction between the flow air, the TEM grid, and the airborne nanoparticles determines the likelihood that the particles will be collected on the grid. As the air passes over the grid, nanoparticles will move randomly around their streamline, and half of them will move toward the grid. As defined in Section 2.6, the root mean square (rms) average distance a particle moves is given by (Hinds, 1999):

$$x_{rms} = \sqrt{2Dt} \tag{2.3}$$

The particle diffusion coefficient is given by

$$D = \frac{kTC_c}{3\pi\eta d_p} \tag{2.2}$$

The Cunningham correction factor is included to account for the departure of small particles, including nanoparticles, from Stokes' law. It is given by the following empirical equation:

$$C_c = 1 + \frac{\lambda}{d}\left[2.34 + 1.05e^{\left(-0.39\frac{d}{\lambda}\right)}\right] \tag{7.A.1}$$

where, λ = mean free path of air molecules, 6.6×10^{-8} m at STP.
An example of using these equations to calculate particle displacement across a TEM grid is included in Section 8.5.1.

REFERENCES

Aitken J. On the number of dust particles in the atmosphere. Trans R Soc Edinb 1888;31:1–19.

Asbach C, Kaminski H, Fissan H, Monz C, Dahmann D, Mulhopt S, Paur HR, Kiesling HJ, Herrmann F, Voetz M, Kuhlbusch TAJ. Comparison of four mobility particle sizers with different time resolution for stationary exposure measurements. J Nanopart Res 2009;11:1593–1609.

Baron PA, Maynard AD, Foley M. Evaluation of aerosol release during the handling of unrefined single walled carbon nanotube material. Publication No. DART-02-191 Rev. 1.1:1–22. Washington (DC): U.S. National Institute for Occupational Safety and Health DHHS (NIOSH); 2003.

Chow JC, Doraiswamy P, Watson JG, Chen LA. Advances in integrated and continuous measurements for particle mass and chemical composition. J Air Waste Manage Assoc 2008;58:141–163.

Fierz M, Kaegi R, Burtscher H. Theoretical and experimental evaluation of a portable electrostatic TEM sampler. Aerosol Sci Technol 2007;41:520–528.

Gangwal S, Brown J, Wang A, Houck K, Dix D, Kavlock R, Hubal E. Informing selection of nanomaterial concentrations for ToxCast in vitro testing based on occupational exposure potential. Environ Health Perspect 2011;119:1539–1546.

Gonzalez D, Nasibulin AG, Baklanov AM, Shandakov SD, Brown DP, Queipo P, Kauppinen EI. A new thermophoretic precipitator for collection of nanometer-sized aerosol particles. Aerosol Sci Technol 2005;39:1064–1071.

Gorbunov B, Priest ND, Muir RB, Jackson PR, Gnewuch H. A novel size-selective airborne particle size fractionating instrument for health risk evaluation. Ann Occup Hyg 2009;53:225–237.

Han JH, Lee EJ, Lee JH, So KP, Lee YH, Bae GN, Lee SB, Ji JH, Cho MH, Yu J. Monitoring multiwalled carbon nanotube exposure in carbon nanotube research facility. Inhal Toxicol 2008;20:741–749.

Hassellov M, Kaegi R. Analysis and characterization of manufactured nanoparticles in aquatic environments. In: Lead JR, Smith E, editors. *Environmental and Human Health Impacts of Nanotechnology*. Chichester: John Wiley & Sons, Ltd.; 2009. p 224.

Hatch T, Warren H, Drinker P. Modified form of the Greenberg-Smith impinger for field use with a study of its operating characteristics. J Ind Hyg 1932;14:301.

Hinds WC. *Aerosol technology—properties, behavior, and measurement of airborne particles*. New York: Wiley-Interscience; 1999.

Jeong C, Evans GJ. Inter-comparison of a fast mobility particle sizer and a scanning mobility particle sizer incorporating an ultrafine water-based condensation particle counter. Aerosol Sci Technol 2009;43:364–373.

Kim JK, Kang MG, Cho HW, Han JH, Chung Y, Rim KT, Yang JS, Kim H, Lee MY. Effect of nano-sized carbon black particles on the lungs and circulatory system by inhalation exposure in rats. Saf Health Work 2011;2:282–289.

Laurent C, Flahaut E, Peigney A. The weight and density of carbon nanotubes versus the number of walls and diameter. Carbon 2010;48:2994–2996.

Lu C, Liu J. Controlling the diameter of carbon nanotubes in chemical vapor deposition method by carbon feeding. J Phys Chem B 2006;110:20254–20257.

Marple VA. History of impactors—the first 110 years. Aerosol Sci Technol 2004; 38:247–292.

May KR. The cascade impactor: an instrument for sampling coarse aerosols. J Sci Instrum 1945;22:187–195.

May KR. An "ultimate" cascade impactor for aerosol assessment. J Aerosol Sci 1975; 6:413–419.

Maynard A, Baron P, Foley M, Shvedova A, Kisin E, Castranova V. Exposure to carbon nanotube material: aerosol release during the handling of unrefined single walled carbon nanotube material. J Toxicol Environ Health A 2004;67:87–107.

Methner MM. Engineering case reports: effectiveness of local exhaust ventilation (LEV) in controlling engineered nanomaterial emissions during reactor cleanout operations. J Occup Environ Hyg 2008;5:D63–D69.

Methner MM, Birch ME, Evans DE, Ku BK, Crouch K, Hoover MD. Case study: identification and characterization of potential sources of worker exposure to carbon nanofibers during polymer composite laboratory operations. J Occup Environ Hyg 2007;4:D125–D130.

Miller A, Frey G, King G, Sunderman C. A handheld electrostatic precipitator for sampling airborne particles and nanoparticles. Aerosol Sci Technol 2010;44:417–427.

Oberdörster G, Oberdörster E, Oberdörster J. Nanotoxicology: an emerging discipline evolving from studies of ultrafine particles. Environ Health Perspect 2005;113:823–839.

Oberdörster G, Stone V, Donaldson K. Toxicology of nanoparticles: a historical perspective. Nanotoxicology 2007a;1:2–25.

Oberdörster G, Oberdörster E, Oberdörster J. Concepts of nanoparticle dose metric and response metric. Environ Health Perspect 2007b;115:A290.

Organisation for Economic Cooperation and Development [OECD]. Preliminary analysis of exposure measurement and exposure mitigation in occupational settings: manufactured nanomaterials Series on the Safety on Manufactured Nanomaterials. OECD Publication No. 8, ENV/JM/MONO(2009)6. Paris: Organisation for Economic Cooperation and Development; 2009.

Pollak LW, Metneiks AL. Intrinsic calibration of the photo-electric condensation nucleus counter 1957 with convergent light-beam. Technical (Scientific) Note No. 9. Dublin: Dublin Institute of Advanced Studies; 1960.

Rose VE, Cohrssen B. *Patty's Industrial Hygiene. Volume 2. Evaluation and Control.* Hoboken (NJ): John Wiley & Sons, Inc.; 2011a.

Rose VE, Cohrssen B, editors. *Patty's Industrial Hygiene. Volume 1. Hazard Recognition.* Hoboken (NJ): John Wiley & Sons, Inc.; 2011b.

Schneider T, Cherrie JW, Vermeulen R, Kromhout H. Dermal exposure assessment. Ann Occup Hyg 2000;44:493–499.

Sinclair D. Optical properties of aerosols. In: *Handbook on Aerosols.* Washington (DC): U.S. Atomic Energy Commission; 1950. p 81–96.

Thayer D, Koehler KA, Marchese A, Volckens J. A personal thermophoretic sampler for airborne nanoparticles. Aerosol Sci Technol 2011;45:734–740.

Tiwary A, Colls JJ. Measurements of atmospheric aerosol size distributions by co-located optical particle counters. J Environ Monit 2004;6:734–739.

Tsai SJ. Potential inhalation exposure and containment efficiency when using hoods for handling nanoparticles. J Nanopart Res 2013;15:1880.

Tsai SJ, Ashter A, Ada E, Mead J, Barry C, Ellenbecker MJ. Airborne nanoparticle release associated with the compounding of nanocomposites using nanoalumina as fillers. Aerosol Air Qual Res 2008;8:160–177.

Tsai SJ, Hofmann M, Hallock M, Ada E, Kong J, Ellenbecker MJ. Characterization and evaluation of nanoparticle release during the synthesis of single-walled and multi-walled carbon nanotubes by chemical vapor deposition. Environ Sci Technol 2009a;43:6017–6023.

Tsai SJ, Ada E, Isaacs J, Ellenbecker MJ. Airborne nanoparticle exposures associated with the manual handling of nanoalumina and nanosilver in fume hoods. J Nanopart Res 2009b;11:147–161.

Tsai SJ, Echevarría-Vega M, Sotiriou G, Santeufemio C, Huang C, Schmidt D, Demokritou P, Ellenbecker M. Evaluation of environmental filtration control of engineered nanoparticles using the Harvard Versatile Engineered Nanomaterial Generation System (VENGES). J Nanopart Res 2012;14:812.

U.S. National Institute for Occupational Safety and Health [NIOSH]. Asbestos by TEM. NIOSH Method 7402, Issue 2. Washington (DC): National Institute for Occupational Safety and Health; 1994.

U.S. National Institute for Occupational Safety and Health [NIOSH]. Diesel particulate matter (as elemental carbon). NIOSH Method 5040, Issue 3. Washington (DC): National Institute for Occupational Safety and Health; 2003.

U.S. National Institute for Occupational Safety and Health [NIOSH]. Draft current intelligence bulletin—occupational exposure to carbon nanotubes and nanofibers. NIOSH Publication No. 2010-XXX. Washington (DC): National Institute for Occupational Safety and Health; 2010.

U.S. National Institute for Occupational Safety and Health [NIOSH]. Occupational exposure to titanium dioxide. NIOSH Publication No. 2011-160. Washington (DC): National Institute for Occupational Safety and Health; 2011.

U.S. National Institute for Occupational Safety and Health [NIOSH]. Current intelligence bulletin 65—occupational exposure to carbon nanotubes and nanofibers. NIOSH Publication No. 2013-145. Washington (DC): National Institute for Occupational Safety and Health; 2013.

U.S. Occupational Safety and Health Administration [OSHA]. Table Z-1 limits for air contaminants. OSHA 29 CFR 1910:1000. Washington (DC): Occupational Safety and Health Administration; 2006.

van de Hulst HC. *Light Scattering by Small Particles*. New York: John Wiley and Sons, Inc.; 1957.

Vesterdal LK, Folkmann JK, Jacobsen NR, Sheykhzade M, Wallin H, Losft S, Moller P. Pulmonary exposure to carbon black nanoparticles and vascular effects. Part Fibre Toxicol 2010;7:33.

Waggoner AP, Charlson RJ. Measurements of aerosol optical parameters. In: Liu BYH, editor. *Fine Particles—Aerosol Generation, Measurement, Sampling and Analysis*. New York: Academic Press; 1976. p 511.

Wang J, Shin WG, Mertler M, Sachweh B, Fissan H, Pui DYH. Measurement of nanoparticle agglomerates by combined measurement of electrical mobility and unipolar charging properties. Aerosol Sci Technol 2010;44:97–108.

Willeke K, Liu BYH. Single particle optical counters: principle and application. In: Liu BYH, editor. *Fine Particles—Aerosol Generation, Measurement, Sampling, and Analysis*. New York: Academic Press; 1976. p 698.

Wittmaack K. In search of the most relevant parameter for quantifying lung inflammatory response to nanoparticle exposure: particle number, surface area, or what? Environ Health Perspect 2007;115:187–194.

8

EXPOSURE CHARACTERIZATION

The concepts of exposure characterization and exposure assessment were discussed in Section 4.1. Looking a little more deeply, exposure assessment is a branch of environmental science that focuses on the processes that take place at the interface between the environment containing the contaminants of interest and the organisms being considered. It looks at the path from the release of an environmental contaminant, through transport, to its effects in a biological system. It tries to measure how much of a contaminant can be absorbed by an exposed target organism, in what form, at what rate, and how much of the absorbed amount is actually available to produce a biological effect. Although the same general concepts apply to other organisms, the overwhelming majority of applications of exposure assessment are concerned with human health, making it an important tool in public health.

The process and science of exposure assessment is to estimate or measure the magnitude, frequency, and duration of exposure to an agent, along with the number and characteristics of the population exposed. A good exposure assessment describes the sources, pathways, routes, and the uncertainties in the assessment and analyzes how an individual or population comes in contact with a contaminant, including quantification of the amount of contact across space and time. The analysis regarding the interaction between a biological system and the exposed contaminants is usually referred as exposure science.

Exposure characterization can be considered as the first step in exposure assessment; it has the much more limited goal of measuring exposures under some

Exposure Assessment and Safety Considerations for Working with Engineered Nanoparticles,
First Edition. Michael J. Ellenbecker and Candace Su-Jung Tsai.
© 2015 John Wiley & Sons, Inc. Published 2015 by John Wiley & Sons, Inc.

specific scenarios. With regard to ENPs, at this time, we are essentially limited to using the tools and techniques described in Chapter 7 to characterize exposures and not to perform a complete exposure assessment. Exposure characterization is used in this chapter as part of the occupational hygiene model, which studies the exposure sources, pathways, magnitudes, characterizations, and engineering solutions to a toxic agent. We will focus on exposures to workers, but the model applies equally as well to the general population.

8.1 EXPOSURE CHARACTERIZATION STEPS

8.1.1 Standard Occupational Hygiene Models

Exposure characterization for workers is performed as part of the standard occupational hygiene (now widely used rather than the older *industrial* hygiene) model for ensuring good worker health. In older references, the model is defined to include three steps, that is, recognition, evaluation, and control, but the model has evolved. Popendorf (2006) defines the new model as having four steps:

Anticipation: the prospective recognition of hazardous conditions based on chemistry, physics, engineering, and toxicology

Recognition: both the detection and identification of hazards or their adverse effects through chemistry, physics, and epidemiology

Evaluation: the quantitative measurement of exposure to environmental hazards and the qualitative interpretation of those hazards

Control: conception, education, design, and implementation of beneficial interventions carried out that reduce, minimize, or eliminate hazardous conditions.

Although exposure assessment, taken broadly as described in the introduction, can be thought of as encompassing anticipation, recognition, and evaluation, the term "quantitative exposure assessment" (Rappaport and Kupper, 2008) is generally synonymous with the evaluation phase.

8.1.2 Exposure Characterization for Nanomaterials

"Exposure Characterization" as defined earlier is a general methodology used by occupational hygienists. When exposure characterization is applied to nanomaterials and nanoparticles, we have found that it requires specific steps to adequately conduct the assessment. Our modified exposure characterization model for exposure activities relevant to nanotechnology requires the following six steps:

1. Anticipate possible exposures to nanomaterials
2. Identify potential exposures and screening
3. Study background issues

4. Monitor and collect contaminants
5. Characterize contaminants
6. Propose engineering solutions

The original "Recognition" phase has been broken into two steps for nanomaterial exposure characterization, that is, step 2, "Identify potential exposures and screening," and step 3, "Study background issues." The "Evaluation" phase is also broken into two steps 4 and 5: 4, "Monitor and collect contaminants" and 5 "Characterize contaminants." Although, strictly speaking, "exposure characterization" does not extend to the "Control" phase, we believe that it is important to "propose engineering solutions" as part of the characterization process; this step should include the consideration of a broad range of solutions for controlling exposures, as discussed in the Chapters 9 and 10.

Exposure can occur in an occupational setting or a broader societal setting and this characterization method can be applied to both. Typically, exposure in an occupational setting can be substantial and it then would require systematic, rigorous characterization methods. The exposure characterization method discussed in this chapter uses examples of its application to occupational settings where exposure to nanoparticles has occurred. Elements required for each characterization step are discussed below.

8.1.2.1 Step 1: Anticipate Possible Exposures to Nanomaterials In this step, we are trying to identify specific nanomaterials used in the workplace which may pose a danger to the workers if they were exposed to them. In general, the anticipation phase relies on both toxicology and epidemiology to identify possible threats, but as of this writing, there are no published epidemiology studies for engineered nanomaterials (ENMs). Thus, for the time being, ENM anticipation will rely on the type of toxicology information discussed in Chapter 5.

8.1.2.2 Step 2: Identify Potential Exposures and Screening Most workplace or occupational settings can be categorized as either a research and development (R&D) workplace or a manufacturing and production workplace. In this step, it is essential that relevant information is gathered and documented to identify potential exposures and follow-up with screening to identify actual exposures and the processes where they are occurring. Important information to be documented includes:

1. Engineering information, which includes production process, scale of production, quantity of materials used, the form of materials used (e.g., powder or liquid), thermal conditions, special engineering design features, and so on.
2. Ventilation system design, which includes general and local exhaust ventilation systems being used, their airflows and velocities, and airflow patterns created in the workplace.
3. Operational factors, which includes operating procedures, duration of tasks, environmental conditions (e.g., temperature and humidity), and worker's motions while performing tasks.

The collected information must be summarized and organized to identify the possible exposure sources and to rank the risk level for the screening process. This screening task involves a rapid investigation of likely exposure levels and can be assisted using a total particle counter such as a DustTrak™ (TSI) or condensation particle counter to provide a range of probable exposure concentrations. The exposures identified as possibly problematic during screening will be subjected to the remaining steps in the characterization process.

8.1.2.3 Step 3: Study Background Issues

8.1.2.3   Step 3: Study Background Issues   One characteristic of airborne nanoparticles that make this category of contaminant different from most others is that there is always a measurable concentration of such particles present in the air, due to natural and industrial sources in our environment. It thus is very important to adequately characterize the background environment in order to determine the contribution of any particular industrial source to the overall nanoparticle concentration. A particle counter or other direct reading instrument will be required to study the background issues during this step.

In addition to the general background nanoparticle concentration, multiple pieces of equipment and production are commonly seen in both R&D and manufacturing workplaces. This may also involve a variety of materials being used in the same room, all of which can be a source of airborne nanoparticles. Identifying the emission source and/or the resulting exposure for some particular nanomaterial when coexisting with other aerosol contaminants can be very challenging. When identifying nanometer-sized contaminants, only trace quantities of such nanomaterials are typically released into the air; such quantities are barely measurable using mass metrics. However, the particle number concentration of such a trace amount of nanomaterials can be substantial, perhaps more than a million particles per cubic meter, which will mix with other contaminants. To deal with the mixture of contaminants in order to identify the nanoparticles of interest will require the following procedures to examine the background contaminants.

Identify Sources of Background Contaminants   Part of the task in step 2 is to sort out the possible processes or the operating conditions that potentially emit contaminants to the room. For example, an electrical cutting saw, a diesel engine, and air conditioner all can emit or carry aerosol contaminants into a workplace. Also, environmental conditions such as open windows, negative pressure in a room, or heating sources can affect the combination of the aerosols present. Such background sources must be identified.

Interpret Background Change   The magnitude of the "uninteresting" background contaminants must be quantified so that it can be separated from the contaminants produced by the process of interest. For this procedure, a real-time instrument is required to detect the releasing source and measure its magnitude. If possible, one should follow-up by setting the proper conditions that can minimize the effect of background contaminants. For example, if an electrical cutting saw is found to significantly release nanoparticles, this equipment should not be operated during the exposure measurements of the source being investigated; if an air conditioner can introduce nanoparticles from outdoors or released from other rooms, consideration should be given to not operating the air conditioner during the measurements.

Use Relative Data The background aerosols usually need to be managed by reducing all controllable contaminant sources to the lowest possible level. However, a background aerosol concentration will still be present except in unusual circumstances, such as performing measurements inside a clean room. Using the relative increase or the ratio of change in the aerosol concentration is helpful to interpret the magnitude of the exposure being evaluated. The starting aerosol concentration in the workplace before any activity or equipment is up and running is used as the "baseline" concentration; this baseline is then subtracted from later measurements to estimate the relative increase in concentration caused by the process being studied. When the baseline concentration is subtracted from the measured concentration, the increased portion of concentration is commonly called the adjusted concentration or corrected concentration.

The measured baseline can be a fixed baseline or a variable baseline. A fixed baseline means one baseline is used to adjust the exposure magnitude throughout the whole process being studied; this is used when the baseline is identified to be stable throughout the studied process, which is the basic and simplest method to obtain the change in concentration. In many cases in actual workplaces, a variable baseline is found, which means that multiple baselines must be used to calculate the adjusted concentrations, or dual instruments are required to obtain the dynamic baseline throughout the process. For example, the released nanoparticles from certain processes or operations in the same room may be found to accumulate over time, and the exposures to be measured are associated with several separate tasks involved in the whole operation. When all extraneous emissions affecting the study workplace are under control, the exposure magnitude associated with each individual task will be the increase in concentration measured before and during the task. Before starting each task, the measured concentration will be the baseline for that task.

8.1.2.4 Step 4: Monitor and Collect Contaminants Steps 2 and 3 help the investigator to figure out possible confounding factors that might mislead the exposure characterization results. When these affecting factors are managed to the extent possible, the tasks of step 4 will be performed. In some cases, when the results of steps 2 and 3 are less than satisfactory, as for example when the background environment consists of a significant amount of contaminants by its nature and cannot be managed; then the tasks for step 4 will need to be adjusted for that situation. The standard tasks discussed in step 4 will apply for common environments where the factors affecting the background can be managed to a certain extent.

To evaluate nanoparticle exposures, it is necessary that two tasks be performed in parallel, as illustrated in Figure 8.1, that is, (1) measurement of the exposure magnitude and (2) collection of aerosol samples for subsequent analysis.

Measurement of Exposure Magnitude Aerosol particles are measured using real-time instruments that provide a continuous monitoring of the fluctuating nanoparticle concentration in the workplace. Several manufacturers provide real-time

Emission from an equipment

(1) Measurement of   (2) Collection of aerosol
exposure magnitude   contaminants

FIGURE 8.1   Illustration of parallel tasks to be performed in step 4. Photos by S. Tsai.

instruments designed for nanoparticle aerosol measurements, which can measure concentration by several different metrics including mass, surface area, number, volume, and particle size distribution. These instruments are described in detail in Section 7.2. Examples of exposure data discussed in this chapter were collected using the TSI 3091 Fast Mobility Particle Size Spectrometer (FMPS), which measures number concentration in the particle diameter range 5.6–560 nm; surface area and volume distributions are calculated by assuming that the counted particles are spherical and mass distribution can be calculated by entering an assumed particle density. The TSI 3936 Scanning Mobility Particle Sizer™ Spectrometer (SMPS) is another instrument that provides measurements similar to the FMPS. Additional instruments are commonly used to supplement measurements by the FMPS or SMPS to provide a more comprehensive evaluation of the aerosol environment. For example, the TSI Aerodynamic Particle Sizer (APS) can be used to measure concentrations for particle aerodynamic diameters between 500 nm and 20 μm. Additional instruments such as the DustTrak, condensation particle counter, or a surface area monitor can be used in parallel to collect additional information.

When performing aerosol measurements, sampling tubing usually is required to transport released aerosols from the source to the real-time instrument. The length of the sampling tubing will vary by the required distance in the physical environment for measurement; typically a 2–3 m length of tubing, consisting of popular materials such as polyvinyl chloride (PVC), plastic (Tygon®), or carbon-impregnated conductive silicone, is connected to the air inlet of the real-time instrument to reach the measurement locations. As discussed in Section 8.6, sample lines should be kept as short as possible to minimize particle loss.

Normalized particle number concentrations are calculated in each of the instrument channels corresponding to a particle diameter, which give the concentration and size distribution profile shown with the particle concentration on the y-axis versus the particle diameter on the x-axis as seen in the example of Figure 8.2.

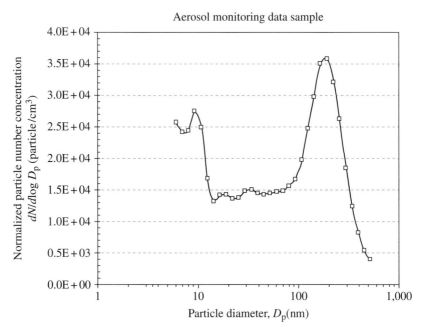

FIGURE 8.2   Example of measured exposure data shown in particle number concentration as a function of particle diameter.

Mixture of Aerosols   Exposure data may have a multiple-peak concentration profile such as the example in Figure 8.2, which may indicate that a mixture of aerosols is present in the measured workplace. Nanoparticles, when released as aerosols, typically are released both as agglomerates and as individual nanoparticles; the proportion of each in a mixture will vary by the particle properties that affect agglomeration and the environmental conditions. A concentration profile with a bimodal peak is typically seen in airborne nanoparticle measurements when agglomeration occurs.

As discussed in step 3, it is necessary to distinguish the nanoparticles being studied from other contaminants in the aerosol mixture. A useful approach is to collect control samples of the material of interest. When the nanoparticles being studied are available in bulk form, a characterization of the bulk nanoparticles can be used to help identify that material's contribution to a measured airborne concentration. When the nanomaterial being studied is not available in its bulk form, the tasks in steps 2 and 3 become essential to provide sufficient background information to interpret the mixture of aerosols.

Collecting Aerosol Samples   Collecting aerosol samples is essential element of nanomaterial exposure characterization in order to fully characterize measured aerosol particles. Since the real-time instruments used to measure aerosol concentrations cannot identify either the morphology or the elemental composition of the measured particles, characterization of collected aerosols is essential in order to provide a complete analysis of the nanoparticles being studied. Particle morphology and elemental analysis can be crucial in identifying and quantifying the sources of aerosol particles in a complex environment. However, selecting proper sampling methods and devices is critical if maximum information is to be obtained.

The particle sampling methods can be categorized into two methods: the "indirect sampling method" and the "direct sampling method," which are described in Section 7.2. Due to the shortcomings of the indirect methods, we have relied on direct methods; practical aspects in using these methods are discussed in Section 8.5.1. Mass sampling methods are not discussed in this section since in most situations we have evaluated, and in the cases discussed in the following texts, ENP mass concentration is not sufficient for analysis. However, mass sampling may sometimes be appropriate if many larger particles are present, for example, when nanoparticles have agglomerated.

8.1.2.5   *Step 5: Characterize Contaminants*   Typically, information on particle size is not sufficient for characterizing nanoparticle aerosols. Information on particle *shape* and *elemental composition* is needed in order to fully characterize the aerosol, for example, to determine the source of the particles and identify the presence of the particular particles being studied. Collected particles will need to be characterized using sophisticated technical instruments such as transmission electron microscopy (TEM), scanning electron microscopy (SEM), energy dispersive spectroscopy (EDS), elemental carbon analysis, and other relevant instruments. A broad range of instruments and tools can be used for characterization and are not limited to those mentioned above.

8.1.2.6   *Step 6: Propose Solutions*   In step 6, the first task is to interpret the exposure characterization data and summarize the results so that proper solutions to address the identified exposure or environmental release can be identified. The exposure data need to be analyzed as discussed in Section 8.3, and linked to the background information collected in step 3. The ultimate goal of performing an exposure characterization is to eliminate those exposures and releases that potentially cause environmental harm and adverse human health effects.

In occupational hygiene, occupational exposure limits (OELs) typically are used as the criteria to judge the acceptability of an exposure or release. However, as discussed in Section 11.1, as of this writing, few if any OELs have been promulgated for ENMs. Setting OELs will require substantial joint efforts of governmental agencies and scientists and the investment of extensive time and financial support. Before adverse human health effects can be known with some certainty and OELs set, it is essential that a protective approach be followed, which includes implementing controls to reduce and prevent exposures. The solutions which might be proposed in this step include engineering control methods (e.g., local exhaust ventilation), process or procedure modification, administrative controls, and workers' work practices, as described in Chapters 9 and 10.

It is essential that the exposure characterization process be repeated after a particular control or engineering solution has been implemented in the workplace being studied. This will involve repeating steps 2–5 to determine the postcontrol exposure levels; some tasks can be simplified or neglected when the outcome can be predicted with certainty.

8.2 EXPOSURE MEASUREMENT STRATEGIES

Workplace exposures can be evaluated using various strategies, that is, single-location measurement, multiple-location measurement, near-field and far-field measurement, and dynamic personal sampling measurement.

8.2.1 Single-Location Measurement

This measurement strategy will take place at one fixed location in the workplace (Fig. 8.3). Concentration measurements using direct-reading instruments and particle sampling will occur at this location throughout the measurement duration. This method is *time* oriented; it is suitable for investigation of the concentration change throughout the study time period. In other words, it does not measure a particular worker's exposure directly, but it can provide an indication of the exposure magnitude if workers are present at the point being monitored.

8.2.2 Multiple-Location Measurement

Under this strategy, measurements will be taken at several locations (Fig. 8.4); the locations will be identified for either concentration measurement or particle sampling or both. This approach is more *source* or *task* oriented. Instruments will be moved to the various sources or tasks of interest to collect data. For instance, to investigate a particular source or task, measurements might be collected at the releasing source, the upwind side or downwind side of the releasing source, the breathing zone (BZ) of a worker adjacent to the source, or from a general room area in the vicinity of the source; these measurements might then be repeated in different locations for different tasks.

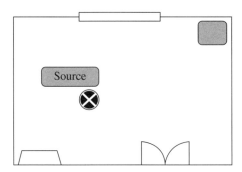

FIGURE 8.3 Illustration of single location measurement; x-marked solid dot is the measuring position.

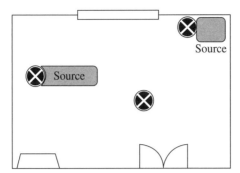

FIGURE 8.4 Illustration of multiple location measurement; x-marked solid dots are the measuring positions.

When a single particle counting instrument is used for multiple location measurements, information is necessarily lost regarding the change in particle concentration over time at any particular location. When two or more identical particle counters are used for measurement, the study can incorporate *time* and *spacial factors* into the measurements. For example, dual instruments can measure particle concentrations upstream and downstream of a filter simultaneously to determine it's collection efficiency; a single instrument would necessarily have to measure them sequentially, and the calculated collection efficiency may be less accurate. As another example, dual instruments can be used to measure particle concentration at two locations that are some distance apart to investigate particle migration from a particular releasing source to different locations in the workplace. A third possibility is to use one instrument to continuously monitor the particle concentration at a fixed location such as a source to determine its fluctuation with the time while relocating the second instrument to various measurement locations.

8.2.3 Near-Field and Far-Field Measurement

The terms "near-field" and "far-field" were originally defined and used when evaluating physical agents such as electromagnetic fields and sound waves. Likewise, the aerosol particle distribution contained in an airflow surrounding a source can be separated into near-field and far-field regions (Fig. 8.5). Researchers have characterized the region between a source and the worker's BZ as the "near field," and this concept has been applied to various toxic substances (Bottini et al., 2003). The far field of a source begins where the near field ends and extends to infinity (theoretically) or the edge of the room (practically).

We can visualize this concept with the hypothetical example of a mechanical operation surrounding a particle-releasing source, where operators are moving about in the immediate vicinity of the machine. In this case, the near-field would be defined by a circle covering the area between the source and the location(s) where the operators normally position themselves to operate the machine. The region beyond this circle would be defined as the far field.

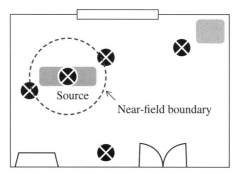

FIGURE 8.5 Illustration of near-field and far-field measurement; x-marked solid dots are the measuring positions.

When using this approach, measurements will be taken at various locations including the boundary region between the near and far fields and close to the source with the goal of determining "typical" near-field and far-field particle concentrations associated with the source. With enough time and measurements, concentrations throughout the far field can be measured systematically in order to obtain a concentration map around the source. When a single instrument is used for this approach, the drawbacks are similar to those described for multiple location measurements; in particular, the *time* factor relative to the particle migration in the region will be hard to be overcome in some situations due to fluctuating particle concentrations.

The three measurement strategies discussed above are measurements at fixed locations that are not linked to a worker's possible movement. When personal exposure data are of interest or concern, dynamic personal sampling measurements must be used.

8.2.4 Dynamic Personal Sampling Measurement

Although currently available real-time instruments are too large to allow for true personal breathing-zone sampling, personal exposure can be evaluated for short-distance movement while a worker is performing a task; such measurements can be of greater interest than measurements at fixed locations. It is possible to measure particle concentration using tubing attached from the instrument to the worker's BZ, which will allow movement over short distances rather than the one fixed location, as shown in Figure 8.6. Practically, the area of movement will be limited due to the limited distance the instrument and tubing can reach. However, since the instruments are designed to be very sensitive, this dynamic measurement can potentially detect small concentration changes due to the worker's motion. Thus, the fluctuation of aerosol concentration could be more intense.

Care should be taken in interpreting very short-term changes in concentration as the worker moves, since the rapid movement of the sampling tube can cause a momentary increase in the measured concentration (perhaps from the re-release of deposited particles from the inside of the tube). For particle characterization of a

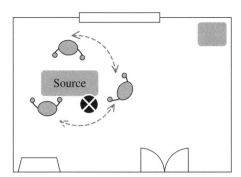

FIGURE 8.6 Illustration of dynamic personal sampling measurement; dashed lines are the motion path.

particular worker's exposure, personal sampling pumps and TEM/SEM samplers can be worn by the workers and used to collect BZ samples; available samplers are discussed in Section 7.2.4 and the sampling methodology is described in Section 8.5.1.

8.3 DATA ANALYSIS AND INTERPRETATION

Data collected using any of the aforementioned approaches will need to be analyzed to interpret the scenario of exposure and environmental release. We will use FMPS data as an example to illustrate typical data calculations and analysis. In ancient times (i.e., before about 2000), most direct-reading aerosol instruments stored collected data for subsequent downloading and processing. With the widespread use of laptop computers, instruments evolved so that now recently developed instruments typically are connected to a laptop during operation and the laptop is programmed to control the instrument and collect and store the data. The raw data stored in the instrument computer is then exported to a readable form such as a text file compatible with MS Excel or other data management system.

In our example, the FMPS measures airborne particle concentrations every second in 32 size channels; this results in a large amount of data being stored in the computer. For example, if the FMPS runs continuously for 8 h, 921,600 data points will be collected and available for analysis. One advantage of the FMPS is that the 1 s sampling time allows for the detection of rapidly changing concentrations, but usually the data can be averaged over a longer time period to facilitate analysis. Some judgment must be used in selecting the averaging time; longer averaging times result in more robust averages and less data to analyze, but at the expense of losing information on concentration changes with time. As a general rule, the averaging time should be consistent with observed fluctuations in the data. For example, if the particle concentration from a source is observed to fluctuate rapidly, the original 1 s data might be analyzed over the entire fluctuation time; if the concentrations are fairly stable, 2 min, 5 min, or a longer averaging time can be used. For the FMPS, the exported data will be "normalized concentrations[1]" of particle number, surface area, volume, and mass concentrations in 32 size channels. The average concentration must be calculated for each individual channel to obtain the concentration profile versus the particle diameter.

Calculated data can be plotted and presented in various ways, for example, concentrations versus particle size (Fig. 8.2), concentrations versus time (discussed in the following and illustrated in Figure 8.10), total concentrations at various locations, and so on, to help interpret the exposure scenarios. Furthermore, near-field and far-field concentration data can be charted using mapping software to view the spatial concentration distribution.

8.4   STATISTICAL ANALYSIS OF DATA

Statistical analysis of collected data may be required to distinguish the differences among various data sets and identify relationships contained within the data. One might ask whether it is necessary for all data to be analyzed statistically. In our opinion, most cases would require it but some would not. Some data sets are obviously different from each other by orders of magnitude, so that statistical analysis to distinguish the difference will not be necessary. For data which fall within the same order of magnitude and for which the differences are not consistent for various particle size ranges, conclusions regarding these data will be difficult to reach without systematic data analysis. Statistical programs such as SAS, SPSS, and so on are commonly available software for statistical data analysis.

This brings us to the question of what statistical analysis methods are most useful for analyzing data as part of a nanoparticle exposure characterization. Figure 8.7 gives an example of FMPS data showing two particle number concentration profiles and the error bars (standard deviation) calculated by averaging the 1 s data over 2 min. Are these two profiles related to each other? In other words, do both curves present comparable concentration size distributions? In some parts of the curves (e.g., at diameters around 100 nm), the concentrations clearly are different, but is one concentration higher than the other at particle sizes around 200 nm, where the error bars overlap? Correlation analysis, such as Pearson correlation, can be used to answer questions regarding the relationship or correlation of two exposure data sets, and the t-test can test the significance of data differences at a particular particle size. These two statistical analysis tools are given below as examples; there are other tools that can be applied to analyze exposure data for particular needs and are not further discussed in this book.

8.4.1   Pearson Correlation

The Pearson correlation (Motulsky, 1995) is used to find a correlation between at least two continuous variables. The correlation coefficient "γ" is a number between +1 and −1 and is interpreted as the magnitude and direction of the association

---

[1] Explanation of normalized concentration is introduced in Appendix 8.A.1.

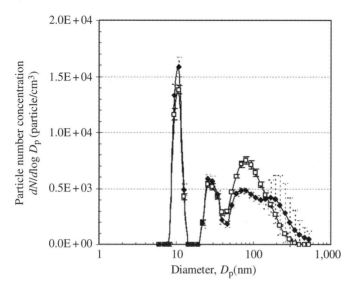

FIGURE 8.7 Particle concentration and size distribution with standard deviation of concentration.

between two variables. The "magnitude" is the strength of the correlation. The closer the correlation is to either +1 or −1, the stronger the correlation. If the correlation is 0 or very close to 0, there is no association between the two variables. The "direction" of the correlation interprets how the two variables are related. A positive γ value means that the two variables have a positive relationship (as one increases, the other also increases), while a negative γ value means that the two variables have a negative relationship (as one increases, the other decreases). In the example shown in Figure 8.7, the particle number concentration as a function of particle diameter is the continuous variable represented by the two curves, and the measured concentrations in the 32 size channels is our estimate of the underlying continuous variables. Pearson correlation can be used to test the correlation between the two 32-element data sets.

8.4.2   *T*-test

The paired samples *t*-test (Motulsky, 1995) is an appropriate test since the FMPS data have the same number of samples from each instrument reading, as long as the same sampling time is selected in each case. The paired samples *t*-test compares the means of two variables. It computes the difference between the two variables for each case, and tests to see if the average difference is significantly different from zero. The *p*-value measures the significance of the difference between the two mean values; a *p*-value less than 0.05 means the compared efficiency profiles are significantly different at the 95% confidence level. For example, to perform a *t*-test for the data in Figure 8.7, one could test the difference between the two sets of concentration data at the 200 nm particle size. The original data used to calculate the mean and standard deviation of the concentration at 200 nm will be the data set for *t*-test. The sample size for the *t*-test will be the number of data points used to calculate the mean and standard deviation; in this case, since 2 min of measurements were used and the FMPS obtains data every second, the sample size is 120.

Other statistical analysis methods can also be used if the number of measurements is sufficient for analysis. Many statistical tests require the underlying distribution to be normally distributed. Measurements in many different settings have found that aerosols size distributions often are log-normally distributed (Hinds, 1999), that is, a plot of the number concentration versus the *log* of the particle diameter follows a normal distribution. It is for this reason that particle size distributions are almost always plotted with a log scale for the diameter, as is done throughout this book. In such case, tests assuming normality can be applied to the logs of particle diameter.

When sampling nanoparticles, however, the measured size distribution usually is not log-normal and frequently has multiple modes. One reason for this is that there usually is more than one source of the measured nanoparticles; another reason is that the measured aerosol frequently consists of primary and agglomerated particles. An option to analyze such multiple-peak data is to separate the merged distribution into several individual log-normally distributed profiles, such as shown in Figure 8.8, using data analysis software such as MATLAB®.

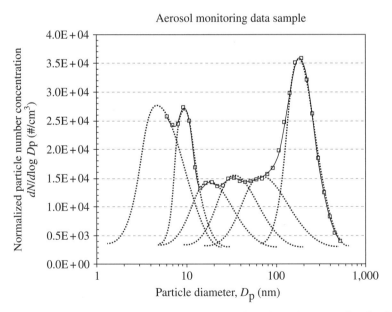

FIGURE 8.8 Separation of concentration profile to individual log-normally distributed profiles.

8.5 PRACTICAL ASPECTS OF AEROSOL SAMPLING AND MICROSCOPY TECHNIQUES

8.5.1 Aerosol Sampling Techniques

The TEM diffusion sampler, the direct sampling tool described in Sections 7.2.5 and 8.1.2, Step 5, collects aerosol nanoparticles through Brownian motion and is a low-cost tool compared to the electrostatic precipitation sampling device that requires a considerable investment. Thus, it is an easy first sampler to use for exposure characterization.

When using this sampler, the air flow and sampling time must be selected with some care. This sampler can be operated at either a low flow (0.1 – 1 L/min) or a medium flow (1 – 5 L/min), depending on the process environment being sampled. The two primary process variables are the length of the process being monitored and the airborne concentration of particle being emitted. First consider airborne concentration. If the FMPS or other automated instrument indicates that the concentration is low, the sample volume (the product of air flow and time) must be high; conversely, a high concentration calls for a lower sample volume. Now consider the length of the process being monitored. If the process is more-or-less continuous, any convenient sampling time can be selected in order to give the desired sample volume; if the process is short-term, however, the sampling time cannot exceed the process time, and the sampler air flow will have to be increased to give the desired sample volume. We frequently use a relatively low sampling air flow of about 0.3 L/min if this will give an adequate sample volume for the process being monitored. This is done because the low filtration velocity increases particle residence time in the filter cassette and enhances the collection of smaller particles by Brownian motion. In other cases, however, we have used sampling flow rates as high as 4–5 L/min in order to get a sufficient sample volume.

We can look more closely at the particle collection mechanisms inside the filter cassette. At our typical pump air flow of 0.3 L/min through a 37 mm diameter filter cassette, the air flows toward the grid at a velocity of 3 mm/s. Air striking the center of the grid will be directed toward the edges of the grid; since the grid has a radius of 1.5 mm, air striking the center of the grid will take approximately 0.5 s to reach the edge. Using the equations from Appendix 7.A.1, we can calculate that a 10 nm diameter particle has a root mean square displacement of 0.23 mm in 0.5 s; since the root mean square displacement is also by definition its standard deviation, 32% of the particles will have moved farther than this distance.

This indicates that a significant number of the nanometer-sized particles in the air flow impinging on the grid will move to the grid surface and be collected. According to this calculation, it indicates that nanoparticles, due to their large Brownian displacement, have a high probability of striking a TEM grid element as they pass by the grid. This particle deposition process has been confirmed experimentally in numerous laboratory and field measurements. Since all particle diameters across the micrometer and nanometer size range will not be collected with equal efficiency by this sampler, the user should not rely on it to represent the whole particle size distribution but rather it should be used for particle morphology and elemental analysis. Several diffusive samplers with different air flows and/or additional samplers such as the ESP or thermal devices described in Section 7.2.5 can be used to collect particles in a broad size range with comparable efficiency.

8.5.2 Microscopy Techniques

Once particles are collected on a grid, TEM analysis can be used to examine individual particle morphology and elemental analysis by X-ray diffraction. In addition, if the collected particles reflect the original aerosol size distribution, the TEM images can be analyzed to determine that particle size distribution, using either manual counting and sizing or an automated counting program such as Gaia Blue (Mirero, Inc.) or Image J (Ferreira and Rasband, 2012). Using such automated tools to supplement instrument measurements is still in a preliminary stage for exposure studies. However, using an image analysis tool has promise to become part of the protocol for exposure characterization and should receive wider use in the future. Results of actual TEM analyses are included in the case studies below. TEM can be supplemented with SEM sampling and analysis if images showing three-dimensional structures provide additional information; an example of this analysis is given in Section 10.6.

8.6   PRACTICAL APPLICATIONS AND LIMITATIONS

8.6.1   Particle Losses

The real-time instruments available in the market now are neither small nor light, making them barely portable and difficult to be placed right at the desired measuring locations. Thus, a sampling tube is needed to transport aerosol particles from the measurement location to the instrument to obtain readings. Particles losses in the sampling tubes have been studied and reported, especially for smaller particles, and adjustment for these losses is important for applications regarding particles in the nanometer size range since losses are the highest due to particle Brownian motion. Particle losses can cause data inaccuracies; it is necessary to understand this issue so that such problems can be avoided. The theory of particle losses is discussed in this section.

Theoretically, particle penetration P (the fraction of particles entering the tube that exit) of a round sampling tube under laminar flow conditions is a function of the dimensionless diffusion deposition parameter, μ (Hinds, 1999):

$$\mu = \frac{DL}{Q} \tag{8.1}$$

where, D=particle diffusion coefficient (defined in Eq. 2.2), m²/s; L=tube length, m; and Q=sampling air flow through the tube, m³/s.

Note that the tube diameter is not included in Equation 8.1; this counterintuitive result is due to the fact that increasing the tube diameter reduces the air velocity in the tube which increases the particle residence time, which would increase diffusional losses, but it also increases the average distance a particle has to diffuse to reach the tube surface. Hinds (1999) gives two different empirical values for the penetration, depending on the value of μ. For $\mu < 0.009$:

$$P = \frac{n_{out}}{n_{in}} = 1 - 5.50\mu^{2/3} + 3.77\mu \tag{8.2}$$

For $\mu < 0.009$:

$$P = 0.819(10^{-11.5\mu^{2/3}}) + 0.0975(10^{-70.1\mu}) \tag{8.3}$$

Several studies (Kumar et al., 2008a, b; Longley et al., 2004) investigated ultrafine particle losses during environmental exposure measurements. Kumar et al. (2008c) evaluated relatively long (5.2–13.4 m) conductive sampling tubes and found that the losses were considerably higher than predicted by Hinds for particles smaller than 20 nm and less than predicted for particles larger than 20 nm.

The effects of bends and elbows of sampling tubes for diffusion of nanoparticles could be significant. Wang et al. found that for air flow in tubing at Reynolds numbers less than 1000, the penetration efficiency in a flow passage with four elbows was up to 40% lower than an equivalent length of straight tubing for particle size of 5–15 nm, which they believe was caused by strong secondary flows in the elbows (Wang et al., 2002). They measured losses in very sharp elbows, however, "specifically the *Swagelock*^{TM} elbows that are found in many aerosol flow systems and instruments." The bends in sampling lines will be much more gradual, so the additional elbow losses should be minimized.

In a research carried out in our lab, which is unpublished, we investigated particle losses in sampling tubing, such as carbon-impregnated conductive silicone tubing and Tygon PVC tubing, with lengths from 2 to 8 m. The total particle loss for particles smaller than 20 nm in the 2 m tubing was found to be 10% or less with two bends in the sampling tube, somewhat higher than predicted by theory. Losses for particles larger than 20 nm were minimal, a result consistent with Kumar et al.'s findings. Losses increased linearly with tubing length, as predicted by theory, but remained negligible for particles larger than about 40 nm for the 8 m length. Interestingly, the conductive tubing, marketed as having less particle loss due to counteracting any electrostatic effects, performed about equally to the Tygon tubing.

Due to the losses of smaller nanoparticles in sampling tubing, and since the degree of particle losses may vary by the tubing materials, and by bends and elbows applied to the tubing, these effects need to be considered when estimating the actual nanoparticle air concentration. However, if the same sampling lines are used, *relative* concentrations at different locations, such as a source and the related BZ, should be useful since the percentage losses will be the same.

Studies performed in a laboratory setting can provide a well-controlled environment to address this particle losses issue, while measurements taken in occupational settings would have limited opportunity to evaluate losses due to the physical conditions in the workplace. The best practical advice is to use the shortest length of tubing that is practical for each particular measurement.

8.6.2   Concentration Measurement versus Particle Samples

When performing real-time nanoparticle measurements, the several available instruments in the market were designed using different theories to estimate particle concentration and size distribution, meaning that the results of measurements using different instruments are usually not identical. The variability of measurement results is a challenge and limitation of using instruments for nanomaterial exposure characterization. In addition to the variability issue, there is no detectable signal from these instrument read, i.e., elemental identification or morphology characterization, that can be used to identify trace amounts of nanoparticles. When the instrument reading cannot provide us detectable data, does this mean no exposure or release occurred in that workplace? For this situation, we cannot reply solely on the direct-reading instrument to evaluate the exposure scenario; here, particle collection for analysis by electron microscopy is more important than the instrument reading to provide evidence of particular particles in the air. Selecting the proper combination of real-time instruments and particle samplers and designing the appropriate measurement plan are crucial to avoiding the misinterpretation of an exposure scenario.

8.7   TYPICAL PRODUCTION PROCESSES

Two examples of production processes involving nanomaterials are introduced in this section, that is, the synthesis of carbon nanotubes (CNTs) and the process to manufacture composites. Each of these production processes is used in one of the case studies that follow, and the cases will refer back to this section regarding the process information.

8.7.1   Synthesis of Carbon Nanotubes

Several techniques have been developed to produce nanotubes in sizeable quantities, including arc discharge, laser ablation, high-pressure carbon monoxide (HiPco), and chemical vapor deposition (CVD). Most of these processes take place in a vacuum or with process gases.

Arc discharge: this is the synthesis method by which CNTs were discovered in 1985 (Iijima, 1991) and the method remains basically the same 40 years after the first discovery and it has been the most widely used method of CNT synthesis. This method produces both single-walled CNTs (SWCNTs) and multiwalled nanotubes (MWCNTs) with lengths up to 50 μm and production yield is up to 30% (Collins and Avouris, 2000).

Laser ablation: the method was developed to synthesize nanotubes a few years after they were first discovered. This method produces SWCNTs and MWCNTs using different catalysts and graphite or a composite of graphite (Guo et al., 1995). The method provides the highest production yield of about 70% and also is the most expensive synthesis method (Collins and Avouris, 2000). By changing the reaction temperature, this method produces SWCNTs with a controllable diameter.

Chemical vapor deposition: this is a common method for the commercial production of CNT and it shows the most promise for industrial-scale production due to its lower unit price. This process is performed in the gas phase at high temperatures above 700°C using a carbon-containing precursor and a catalyst, and CNTs typically are formed inside the reaction chamber. The production yields of this process vary by the operating conditions.

8.7.2 Composite Manufacture

A composite is made from at least two discrete substances, and this term is usually used for various industrially manufactured composites such as reinforced plastics. To form a polymer composite, reinforcing materials such as carbon black, fibers, or a metal oxide are added as fillers to a polymer to form a polymeric matrix that will provide higher strength and other designated property enhancements compared to the polymer alone.

In the past, the fillers were usually in the micrometer size range, and during the last decade, the industry has been downsizing to the nanometer range. Such a material using nanoparticles as reinforcing agents is termed a nanocomposite. Nanocomposites have the potential to be implemented as new high-strength replacements for traditional composites. Fillers used for manufacturing nanocomposites include alumina, carbon black, silica, talc, calcium carbonate, layered silicates (nanoclays), and, recently, silver and engineered nanoparticles such as CNTs.

Although the low filler loadings (<10% by weight) in nanocomposites permit retention of flexibility and impact properties, properties are highly dependent on dispersion of the primary filler particles through the polymer matrix. If clusters are avoided, then the small diameter of the particles themselves will contribute a very high interfacial surface area and in theory increase the mechanical properties of the polymer (McCarrie and Winter, 2003).

An example of nanocomposite compounding using a twin screw extruder (TSE) and feeders such as twin screw feeders to be placed on the material loading port to feed materials into the extruder is shown in Figure 8.9. A typical operation of compounding process includes five time periods as illustrated in Figure 8.10 (Tsai et al., 2008a). During phase I, the warm up period, the TSE is warmed up from room temperature to above 200°C. Phase II is for setup and calibration, when nanoparticles are loaded from the twin screw feeder to calibrate the feeding rate. Phase III is for compounding of virgin polymer without adding nanoparticles, where the virgin materials are used as a reference; this phase can be neglected if the reference is not needed. Phases IV and V are for compounding of nanocomposites, when nanoparticles are

FIGURE 8.9   Equipment and setup of composite manufacturing. Photo by S. Tsai.

I. Warm up	II. Set up	III. Feed polymer	IV. and V. Compound with fillers	
		Process flow direction	→	Nanocomposite product

FIGURE 8.10   Illustration of the time sequence during compounding and the timing of particle measurement. Roman numerals indicate the phases of a typical experiment. Reprinted with permission from Tsai et al. (2008a). Copyright 2008 the Taiwan Association for Aerosol Research.

fed by a twin screw feeder into the extruder in parallel with feeding polymer pellets through feeding port. For the case study described here, different ratios of nanoparticles to the polymer were employed for phases IV and V, that is, 2 and 5% of nanoparticles (aluminum oxide) by weight.

The temporal patterns of total number concentration and median diameter are presented in Figure 8.11; some results from this research will be discussed in the following case study, Section 8.10. This long-term monitoring shows total number concentration measured from the beginning of warm up to the end of the operation following the time phases shown in Figure 8.10.

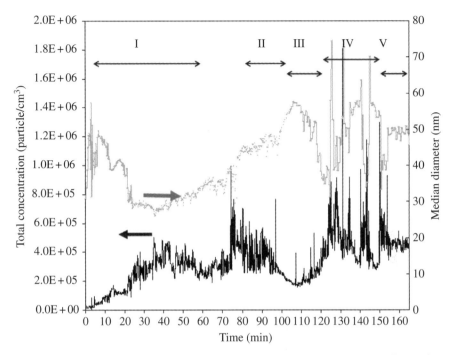

FIGURE 8.11 Total number concentration and particle median diameter at different time periods and operations of experiments. This compounding process was operated using twin screw feeder feeding in the primary feeding port (the first port). Black curve is concentration profile; the gray curve is particle median diameter. Reprinted with permission from Tsai et al. (2008a). Copyright 2008 the Taiwan Association for Aerosol Research.

8.8 CASE STUDY: MANUAL HANDLING OF NANOPARTICLES

Handling nanoparticles in the powder form is a typical practice for many work activities in both research and production settings. Bulk materials of nanoparticles contain agglomerates, small or large, and individual particles in the nanometer size range, and when released as aerosols will not be visible to workers. How high could exposures to such invisible nanoparticle aerosols be when released during manual manipulation? This case illustrates health and safety issues associated with the manual handling of nanoparticle powders.

Case theme: Manual transferring or pouring nanoparticle powder for material weighing in a fume hood.

Equipment: 100 g or less of nanoparticles, beakers, spatulas, fume hood.

Process: (i) Transfer of powder by pouring from one beaker to another. (ii) Transfer of powder from one beaker to another by means of a spatula.

Investigation: Nanoparticle exposure to the worker, nanoparticle release surrounding the transferring beakers.

8.8.1 Materials and Conditions

Aluminum oxide[2] (nanoalumina) and silver[3] (nanosilver) nanoparticles were handled to demonstrate weighing tasks; 15 or 100 g of nanoparticles were transferred using a spatula or poured from beaker to beaker. The nanoalumina particles used here appear roughly spherical in shape, have a reported density of 3600 kg/m³, and a primary particle size ranging from 27 to 56 nm; when dried, these particles formed agglomerates in the bulk material with a nominal size of approximately 200 nm. The nanosilver particles used here were manufactured as a grayish-white powder, appear roughly spherical in shape, and have a density of 10,500 kg/m³ and an average particle size of 60 nm.

Exposure and release from particle handling was evaluated by performing the handling operations in a conventional hood, which has a constant air flow, with no special features to help control the hood face velocity, so that the hood face velocity varies inversely with the height of the sash opening (see Section 9.1.2). The conventional hood used here has a full open sash face dimension of 62 cm (24 in.) high × 130 cm (51 in.) wide. Measurements taken in this hood were taken at conditions of relative humidity in the range 48–52%.

8.8.2 Particle Handling

Particle handling was performed in the hood as shown in Figure 8.12 (Tsai et al., 2009a). Beakers were placed in the hood on an analytical balance. The transferring task was performed by using a spatula to transfer nanoparticles from one beaker to another (Fig. 8.12a); 0.7–1.2 g of nanoalumina was loaded into the open top of the beaker during each spatula transfer. For the pouring task (Fig. 8.12b), nanoparticles were poured directly from one beaker into a second beaker at the center of the open hood face, so that the feeding and receiving beakers were adjacent to each other at the open edge. When handling 100 g of nanoalumina, particles were transferred between 400 ml beakers. Pouring 100 g of nanoalumina took about 1 min and transferring took about 4.5 min. For 15 g nanoalumina handling, 100 ml beakers were used because the higher-density particles had less volume. Pouring 15 g of nanoalumina took about 20 s and transferring took about 90 s.

8.8.3 Measurements

Measurements were taken at the researcher's BZ and the area surrounding the particle transfer, called the source location (Figs 8.12c, d, 8.13 "A, B"). BZ concentration was measured near the researcher's nose (Figs 8.12e and 8.13 "C"). Nanopowders were handled in the hood on the work surface, 15 cm (6 in.) back from the sill. Particle

---

[2] Also called nanoalumina, the NPs used here were grade Al-015-003-025, manufactured using physical vapor synthesis (PVS) by Nanophase Technologies Corporation, Romeoville, IL, http://www.nanophase.com/technology/

[3] Also called nanosilver, the NPs used in this study were engineered for increased electrical conductivity in low-temperature processing scenarios and manufactured by NanoDynamics Inc., Buffalo, NY.

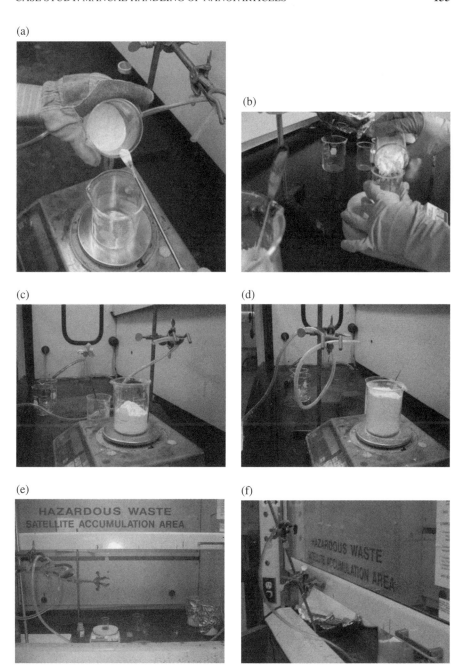

FIGURE 8.12 Manual handling experimental setup and locations of measurement. (a) Manual transferring nanoparticle powder; (b) manual pouring nanoparticle powder; (c) source downstream side measurement; (d) source upstream side measurement; (e) middle sash position, 15 g nanosilver transferring; and (f) low sash position, breathing zone measurement. (a) and (b): reprinted with permission from Tsai et al. (2009a). Copyright 2009 *Journal of Nanoparticle Research*. (c)–(f): photos by S. Tsai.

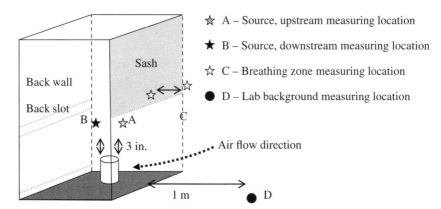

FIGURE 8.13 Schematic showing different measurement locations. Reprinted with permission from Tsai et al. (2009a). Copyright 2009 *Journal of Nanoparticle Research.*

concentrations were measured at the upstream and downstream sides of the releasing source. Upstream and downstream measuring locations (Figs 8.12c, d and 8.13) were 8 cm (3 in.) vertically above the beaker at the upstream and downstream edge, respectively. For the conventional hood, sash locations for measurement were full open, half open (Fig. 8.12e), and low chest height (Fig. 8.12f), which corresponded to 62 cm (24 in.), 44 cm (17 in.), and 16.5 cm (6.5 in.) open sash, respectively.

The concentrations of airborne nanoparticles were measured using the FMPS, described in Section 7.2, for particle diameters from 5.6 to 560 nm. Particle concentration and size distribution were recorded every second. Normalized particle number concentrations were calculated in each size channel based on the average concentration during each measuring time period. The concentration measured before handling particles was used as the baseline for subtraction from the particle concentration. A 3 m length of Tygon tubing was connected to the air inlet of the FMPS to reach the measurement locations.

8.8.4   Aerosol Particle Characterization

Aerosol particles were collected using the diffusive sampler as described in Section 7.2.4 and shown in Figure 7.16 for further analysis to obtain particle morphology and elemental composition. TEM copper grids (400 mesh with a titanium dioxide film) were taped onto 47 mm diameter polycarbonate membrane filters (0.2 µm pore size). Air flow was provided by a personal sampling pump at a rate of 0.3 L/min.

Scanning Electron Microscope[4] (SEM) images of particles collected on the membrane filters were obtained by coating each sample with a thin layer of gold using a Denton Vacuum Desk IV cold sputter unit and then analyzing the sample at an accelerating voltage of 1–15 kV of electron beam energy. Scanning transmission

---

[4] Images of the samples were taken using a JSM-7401 F Field emission scanning electron microscope (JEOL, Peabody, MA)

electron microscope (STEM) images of particles collected on the TEM grids were obtained using an electron microscope operated at an accelerating voltage of 20 kV. Elemental analysis was performed using an EDS attachment of the SEM (EDAX, Mahwah, NJ) with a primary electron beam excitation energy of 10 kV.

8.8.5 Results

8.8.5.1 Effects of Nanopowder Quantity and Location on Particle Release The white nanoalumina particles visibly contaminated the weighing balance, hood surface, and tools placed nearby the balance, as seen in Figure 8.12b, c, and d, which is a visible evidence that some particles had settled onto the hood work surface while following the air flow toward the hood back slots. Since we know that individual nanometer-sized particles are not visible, what quantity of such nanometer-sized particle must have been released, either as individual nanoparticles or as micrometer-sized agglomerates, to create visible surface contamination?

Airborne nanoparticle concentrations were measured during the weighing tests using the FMPS. One set of measurements, shown in Figure 8.14, documents the magnitude of the particle release at three locations, that is, source upstream side, source downstream side, and worker's BZ, while transferring 100 g of nanoalumina particles. The BZ concentration is lower compared to measurements surrounding the source, and the downstream side of the source had the highest concentration of larger nanoparticles, indicating significant particle release. The upstream and BZ measurements indicate that the further the distance away from the source in the reverse airflow direction

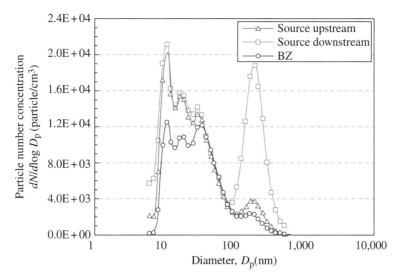

FIGURE 8.14 Particle concentration and size distribution at three locations during transferring 100 g nanoalumina in the conventional hood. Reprinted with permission from Tsai et al. (2009a). Copyright 2009 *Journal of Nanoparticle Research.*

from the handling spot, the lower will be the particle concentration. This is consistent with the theory of particle transport and diffusion.

8.8.5.2 *Effect of Working Conditions on BZ Concentration*   Measurements taken during this study showed that particle number concentration measured at the researcher's BZ increased significantly during 100 g nanoalumina particle handling using the conventional hood. The BZ exposure data before and during transferring and pouring nanoalumina particles are plotted in Figure 8.15a and b, respectively. Most BZ exposure data were above the baseline before the beginning of the experiment, indicated by "BZ before expt," indicating that the concentration increase is due to the nanoparticle handling and is evidence of exposure to the handling worker.

The magnitude of nanoparticle exposure can be quantified by subtracting the baseline data in each particle size channel in order to obtain the concentration increase contributed by the particle handling task. The resulting increases in particle concentrations (commonly called adjusted or corrected concentration) as a function of particle diameter are plotted in Figure 8.15c and d. For a few channels ($d_p < 10$ nm) where exposure concentrations were lower than the baseline, the adjusted concentrations were negative values and are not shown in the figures. The increase in particle concentration varied by particle size, sash location, and handling method (transferring or pouring); the adjusted concentration ranged from several thousand to over ten thousand particles per cubic centimeter at the researcher's BZ. Looking at measurements at the three face velocities, all size distributions showed likely bimodal curves with one peak at particle size of approximately 200 nm, which was the most common size of agglomerated nanoalumina in the bulk material.

The exposure concentration increase using the spatula transferring method showed a similar size distribution and magnitude at all three velocities (Fig. 8.15c). While pouring (Fig. 8.15d), more particles were carried out of the hood during handling at the highest face velocity of 1.0 m/s when the sash was lowered to the researcher's low chest height. The increase in particle number concentration measured at the BZ during pouring was as high as 13,000 particles/cm^3.

Airborne particles detected at the BZ were also collected on copper tapes using the diffusive sampler (Fig. 7.16) during handling 100 and 15 g of nanoalumina; collected particles analyzed using SEM are shown in Figures 8.16a and b, respectively. Particles were confirmed to be alumina by elemental composition, as shown by the elevated "AlK" (aluminum/potassium) energy peak indicated by the arrow mark on the EDS spectrum. Particle sizes were found in a broad range from submicrometer to several micrometers. Many nanoparticles in the form of large agglomerates were collected at the worker's BZ during handling 100 g of nanoalumina as can be seen in Figure 8.16a, while fewer nanoalumina particles were collected during handling 15 g of nanoalumina, as shown in Figure 8.16b.

In many laboratory settings, nanoparticles are usually handled in quantities as small as a few grams to micrograms. The tests using 15 g of nanopowder were included to investigate this scenario. When handling 15 g of nanoalumina in the conventional

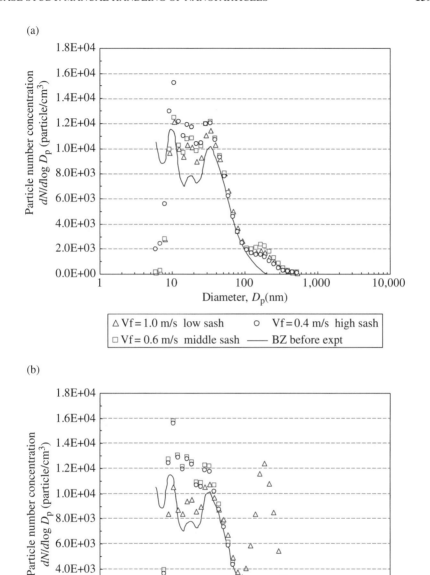

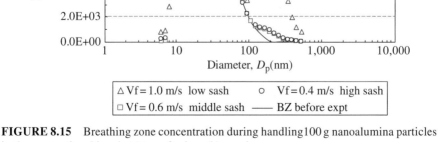

FIGURE 8.15 Breathing zone concentration during handling 100 g nanoalumina particles in the conventional hood: (a) transferring, (b) pouring,

(c)

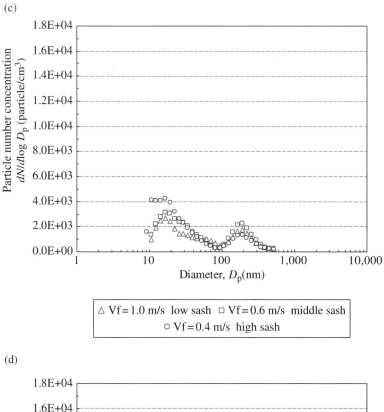

(d)

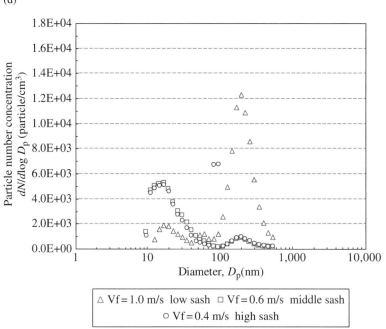

FIGURE 8.15 (*Cont'd*) (c) concentration increase during transferring, and (d) concentration increase during pouring. *Y*-axis: Relative normalized particle number concentration calculated using measured concentration subtracting average background concentration. *X*-axis: diameter of the average particle size in each channel of the FMPS. Reprinted with permission from Tsai et al. (2009a). Copyright 2009 *Journal of Nanoparticle Research*.

(a)

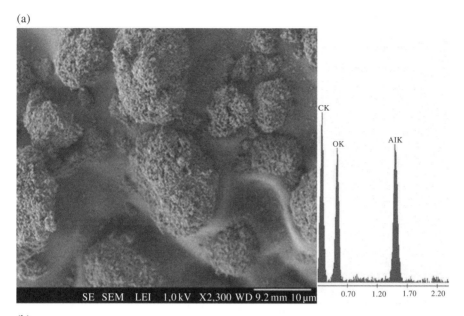

(b)

FIGURE 8.16   SEM photos of nanoalumina particles: (a) handling 100 g and (b) handling 15 g. Reprinted with permission from Tsai et al. (2009a). Copyright 2009 *Journal of Nanoparticle Research*.

hood, adjusted concentrations at the BZ were consistently lower compared to handling of 100 g of nanoalumina particles, as can be seen in Figure 8.17. For both transferring and pouring, the highest exposure occurred at the face velocity of 1.0 m/s when the concentration peaked at 500–600 particles/cm^3 above background levels.

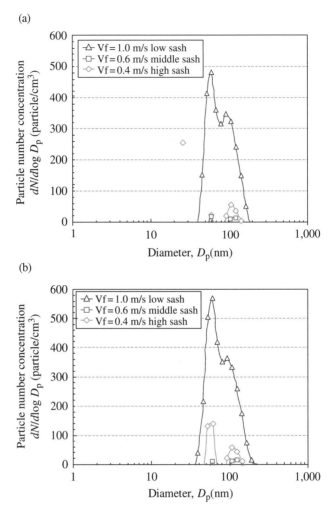

FIGURE 8.17 Increase in the breathing zone concentration during handling 15 g nanoalumina particles in the conventional hood: (a) transferring and (b) pouring. Reprinted with permission from Tsai et al. (2009a) Copyright 2009 *Journal of Nanoparticle Research.*

8.8.5.3 Effect of Nanomaterial Type

The results from handling nanoalumina particles showed that high exposure could potentially occur during the weighing of a nanopowder. How would the particle release and exposure be affected when different types of nanomaterials are handled? Four measurements taken at the downstream side of the source while handling nanoalumina particles were compared to measurements taken during handling nanosilver particles to investigate the effect of material on particle release. The measured concentrations while handling nanoalumina and nanosilver particles are shown in Figures 8.18 and 8.19, respectively. As shown in Figure 8.18, significant numbers of nanoalumina agglomerates peaking at 200 nm appeared on the downstream side for all measurements. However, nanosilver agglomerates peaking at

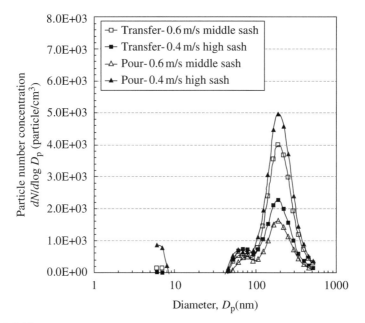

FIGURE 8.18 Particle concentration increase at the downstream side during handling 15 g nanoalumina in the conventional hood. Reprinted with permission from Tsai et al. (2009a). Copyright 2009 *Journal of Nanoparticle Research*.

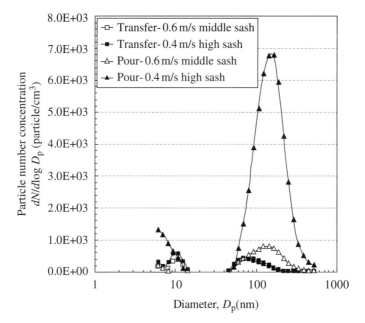

FIGURE 8.19 Particle concentration increase at the downstream side during handling 15 g nanosilver in the conventional hood. Reprinted with permission from Tsai et al. (2009a). Copyright 2009 *Journal of Nanoparticle Research*.

FIGURE 8.20 TEM images of nanosilver: (a) agglomerate at low magnification; (b) agglomerate at high magnification; and (c) bulk material. Reprinted with permission from Tsai et al. (2009a). Copyright 2009 *Journal of Nanoparticle Research.*

150 nm were detected on just one of four measurements (Fig. 8.19). FMPS concentration data collected during handling 15 g nanosilver showed that only one measurement was elevated, and few nanosilver agglomerates were found among particles collected at the researcher's BZ as can be seen in Figures 8.20a and b. Since collected silver particles were several micrometers in diameter and no particles in the nanometer size range were found on the filters, most likely the bulk nanosilver material undergoes intense agglomeration as seen in Figure 8.20c.

8.8.6 Discussion

8.8.6.1 Effect of Material Quantity and Type on Particle Release These tests demonstrate the likelihood of particle release and exposure during the handling of nanoparticles and confirm the logical finding that handling larger quantities of nanopowders increased the airborne release of nanoparticles, leading to a greater number escaping the hood and reaching the researcher's BZ. The larger quantity handled (100 g) is more than six times greater than the smaller amount (15 g), but the ratio of the number of particles released is much greater than a factor of six. More studies are required to correlate particle release with quantity handled with any degree of certainty.

Considering the type of nanoparticle handled, the reason that more nanoalumina than nanosilver particles were detected inside the hood while pouring is likely due to different flow patterns of nanoalumina and nanosilver when dumping bulk nanoparticle agglomerates into the beaker. In addition, due to differences in density, the volume of nanoalumina used is about three times the volume of nanosilver for the 15 g handling task, yielding a larger number of nanoalumina particles being handled than nanosilver. The low-density nanoalumina agglomerates are less affected by gravity and thus settle more slowly in bulk flow than the bulk nanosilver particles. The displaced air stream flowing upward inside the beaker while pouring nanoalumina particles behaved differently than when pouring nanosilver. Upward-flowing air easily penetrated the plug flow of the falling nanoalumina nanoparticles, whereas the denser nanosilver particles fell as one mass, making air penetration more difficult. This resulted in many more nanoalumina particles being entrained into the displaced air stream compared to the number of nanosilver particles.

8.8.7   The Challenge and Brainstorming

As she looked at a sealed bottle of nanoparticle powder, Christina thought about the possibility that she could be exposed to nanoparticles while performing her work. Her first task was to measure 45 g of titanium dioxide (TiO_2) dry nanoparticle powder and pour the powder into a liquid to prepare the TiO_2 liquid suspension for the next step, a coating process. She pondered what steps she could take to avoid exposure. In the laboratory where she worked, there was one conventional hood and space on a bench top that she could use. The laboratory room was served by general ventilation from the building HVAC system.

Christina thought that she likely would be exposed to some level of TiO_2 nanoparticle if she did weighing in the conventional hood. Is she correct? She also considered doing the weighing on the bench top. What other factors she should consider to perform her task?

8.8.8   Study Questions

1. What types of local exhaust ventilation system other than the conventional hood are available in occupational or laboratory settings?
2. What factors when using the conventional hood could cause nanometer-sized contaminants to escape the hood to reach the worker's BZ?
3. What would be the difference when micrometer-sized particles compared to nanometer-sized particles are handled in the hood? What physical properties of particles are relevant to this?

4. When a sophisticated instrument is unavailable to detect airborne nanoparticle during the handling task, can any alternative method be used to judge evidence of exposure?

8.9   CASE STUDY: SYNTHESIS OF CARBON NANOTUBES

Several techniques have been developed to produce nanotubes in sizeable quantities, including arc discharge, laser ablation, high-pressure carbon monoxide (HiPco), and CVD. Most of these processes take place in vacuum or with process gases. CVD, considered a combustion process, is a rapid chemical reaction between substances that is usually accompanied by generation of heat and light in the form of flame. CVD growth of CNTs can occur in a vacuum or at atmospheric pressure. Large quantities of nanotubes can be synthesized by these methods; advances in catalysis and continuous growth processes are making CNTs more commercially viable. However, the yield of these methods varies by technique and is not high. The unconverted substances will form carbonaceous materials mixed with CNT fragments to be exhausted as an emission. Concerns were raised about emissions from CNT synthesis causing exposure and potential environmental impact.

Case theme: Synthesis of CNTs using the CVD process on a laboratory scale.

Operation: CNTs were synthesized in a furnace at temperature of 800°C, and exhaust from the heated furnace was emitted into a fume hood.

Investigation: Morphology and magnitude of particles in the furnace exhaust air.

8.9.1   Materials and Synthesis

This case study evaluates the emissions from MWCNT production through continuous catalyst injection. The production was performed for two scenarios: growing MWCNTs on a substrate and having no substrate for the growth. For the no-substrate scenario, two different conditions, high and low injector temperature, were tested to evaluate the effect of temperature on released nanoparticles during synthesis. The notation used to identify the different experimental conditions is listed in Table 8.1.

All experiments were performed using a typical laboratory CVD setup consisting of a 2.5 cm × 60 cm (1 in. × 24 in.) fused silica cylindrical reactor chamber heated by a clamshell furnace located in a laboratory constant velocity fume

TABLE 8.1   Experimental notation

Multiwalled CNTs (MWCNT) production	
Low injector temperature	A: use substrate (L temp-A)
	B: no substrate (L temp-B)
High injector temperature	B: no substrate (H temp-B)

hood. Aerosol-assisted CVD was used to generate MWCNTs at high yield following the procedure of Xiang et al. (2007). In this approach, both the catalyst and the carbon feedstock were continuously introduced into the reaction zone during the 20 min CVD process by evaporating a solution of ferrocene and cyclohexane at a feeding rate of 10 ml/h from a heated nozzle. The resulting CNTs grew on the reactor walls and Si/SiO$_2$ substrates and exhibit a forest-like alignment.

Varying the temperature of the injector nozzle affected the morphology of the obtained nanotube films; at a low nozzle temperature of approximately 100°C, forest-like MWCNT structures of 600 μm height were synthesized whereas increasing the nozzle temperature resulted in carpet-like structures. The high reaction temperature was 800°C and gas streams of 500 scm^3/min Ar and 50 scm^3/min H$_2$ were used. Different growth patterns can be obtained by varying the application rates of these materials. The continuous introduction of catalyst and carbon feedstock in the case of MWCNT growth creates high-density films for structural applications.

8.9.2   Measurement

Particle counters, i.e., the FMPS and APS, described in Section 7.2, were used to monitor particle release. Measurements were taken at the background, BZ, and source locations as illustrated in Figure 8.21a (Tsai et al., 2009b). Background particle concentration was measured 1 m (3 ft) from the fume hood to represent the concentration in the room. The BZ concentration was measured in the researcher's working area which was about 0.5 m (18 in.) from the hood as shown in Figure 8.21b. Also shown is the sampling location of the FMPS and APS at the particle release source, designated the source location, which was inside the fume hood. The source location was defined as the exhaust of the reactor inside the fume hood. Since the exhaust gas flow (2 L/min) was less than the required FMPS (10 L/min) and APS (5 L/min) sample air flow, the incoming exhaust stream was diluted with air from the fume hood as shown in Figure 8.21b.

The exhaust normally went through a cellulose acetate filter, which was not used for these experiments in order to measure actual particles released from the furnace. Background concentration in the room was recorded before, during, and after each operation. Source concentration was recorded before and during the synthesis of CNTs.

The nanoparticle aerosol filter sampler as described in Section 7.2.4 and shown in Figure 7.16 was used to collect aerosol nanoparticles in these experiments. The sampling location is shown in Figure 8.21b. TEM copper grids (SPI 400 mesh with a Formvar/carbon film) were taped onto 47 mm diameter polycarbonate membrane filters (0.2 μm pore size). Fiber backing filters were used to support the polycarbonate filters. Air was drawn through the filters at 0.3 L/min using a calibrated personal sampling pump, and aerosol particles were deposited on the grid via Brownian diffusion. TEM sample analysis was performed as described in Section 8.4.

(a)

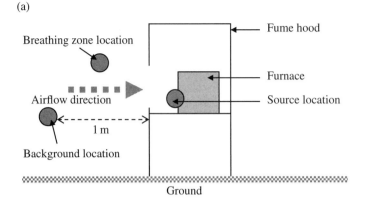

(b)

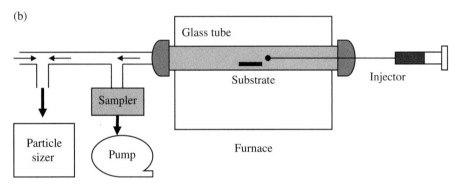

FIGURE 8.21 Chemical vapor deposition (CVD) furnace. (a) Illustration of measuring locations and (b) illustration of process diagram. Reprinted with permission from Tsai et al. (2009b). Copyright 2009 *American Chemical Society.*

8.9.3 Results

8.9.3.1 Comparison of Particle Concentrations at Different Locations Typical results of the aerosol concentration measurements at the furnace exhaust during synthesis of MWCNTs are shown in Figure 8.22 (furnace exhaust), Figure 8.23 (room background), and Table 8.2. The background concentrations in the nanoparticle size range were all very low compared to those at the source, as shown in Table 8.2. Monitoring for the experimental combination of high injector temperature using a substrate was not performed because it resulted in a low production yield, and thus this condition was not of interest for the manufacturing process. In Figures 8.22 and 8.23, the primary y-axis on the left is the scale for particle concentrations in the complete size range (5–20,000 nm) which are shown by the solid line curves; the secondary y-axis on the right is the scale for particle concentrations for the larger size range 550–20,000 nm which are shown by the dotted line curves.

The particles released at the source during synthesis were found at a high concentration (Fig. 8.21), more than 4×10^6 particles/cm^3; the measurements at the

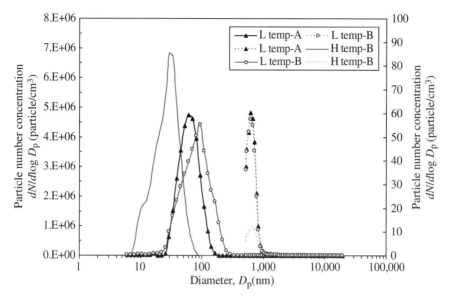

FIGURE 8.22 Particle number concentration and size distribution at source location. L temp-A: low temperature, use substrate; L temp-B: low temperature, no substrate; and H temp-B: high temperature, no substrate. Reprinted with permission from Tsai et al. (2009b). Copyright 2009 *American Chemical Society*.

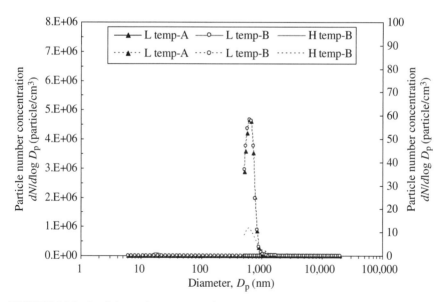

FIGURE 8.23 Particle number concentration and size distribution at background location. L temp-A: low temperature, use substrate; L temp-B: low temperature, no substrate; and H temp-B: high temperature, no substrate. Reprinted with permission from Tsai et al. (2009b). Copyright 2009 *American Chemical Society*.

TABLE 8.2   **Total particle concentrations for different conditions**

| | Total concentration (particle/cm³) | | | |
| | FMPS 5–560 nm | | APS 500–20,000 nm | |
	Source	BG	Source	BG
L temp-A	1,985,000	5,200	19	18
L temp-B	2,313,900	3,840	18	19
H temp-B	3,141,000	18,690	4	5

BG, background location

background, however, were low and close to the baseline (Fig. 8.23). The released particle size distributions varied by the operating condition, but most particles were less than 300 nm in diameter. The particle concentrations and size distribution in the range above 550 nm were very similar at the source and the background, indicating that no aerosol particles above one-half micrometer in diameter were released at the source during synthesis.

The total particle concentrations for the measurements plotted in Figures 8.22 and 8.23 are shown in Table 8.2. The numbers in Table 8.2 are the actual FMPS and APS data, which were not adjusted, thus, these data were affected by two experimental conditions. First, approximately 15% of the furnace exhaust was diverted to the filter sample, so that the number of particles reaching the FMPS and APS was reduced by this fraction. Second, the furnace air was diluted with room air at a rate of approximately 4 to 1 and 5 to 2 for FMPS and APS, respectively, so the actual furnace concentrations would have been higher than the values reported in Table 8.2 by the same factors. Thus, the values in Table 8.2 only reflect relative particle concentrations under different conditions, not absolute ones. The concentration measured at the background location was not diluted and it was extremely low compared to the source. The total concentrations for particle sizes greater than 500 nm at both the source and background are very similar, and background levels outside the fume hood during synthesis were identical to background levels before processing began, indicating no release of particles from the hood.

8.9.3.2   *Comparison of Substrate Use and Temperature*   Three conditions were evaluated while measuring particle release during synthesis of MWCNTs. In the case of low injector temperature, two scenarios were chosen: (1) (L-temp A): a substrate was placed inside the reaction chamber to collect the CNTs during growth and (2) (L-temp B): no substrate for MWCNT growth was used. For the case of high injector temperature (H-temp B), no substrate was used. The quantity and size of particles below 550 nm that were released during synthesis was affected by the operating conditions. Particles released at the source when using a substrate were in the size range of 25–100 nm and the particle release when the substrate was not present yielded a similar but broader particle size distribution from 20 to 200 nm.

The variation in temperature of the injector needle results in significant differences in the particle concentration and size distribution of released particles at the source as seen on the solid curves "L temp-B" and "H temp-B" in Figure 8.22. Particles released at the high temperature condition were in a smaller size range of 7–90 nm compared to the particles released at low temperature condition with a size range of 20–200 nm. In addition, the highest particle concentration measured at the high temperature condition approached 7×10^6 particles/cm³. The particle concentration above 500 nm size showed a much lower value of 10 particles/cm³ peaking at 700 nm at the condition of high injector temperature compared to the other conditions. However, the total concentrations at the source and the background locations are similar in all cases for particles above 500 nm.

8.9.3.3 *Characterization of Aerosol Particles by TEM*   Particles released in the furnace exhaust were collected using the aerosol filter sampler. The particles collected on grids under the different operating conditions were characterized by TEM as shown in Figure 8.24. For low-temperature injector conditions, the particle concentration was higher during no-substrate synthesis as shown in Figure 8.22 and Table 8.2, and indeed the TEM images showed a higher density as can be seen in Figure 8.24a and b. In addition, for the low injector temperature, the collected particles were predominantly clusters of spherical shape with some individual nanoparticles, as well as MWCNTs. The individual spherical nanoparticles were as small as 20 nm, and the size of clusters had a broader distribution but were no larger than 300 nm as can be seen in Figure 8.24a and b. This result is consistent with the measurement of particle concentration shown in Figure 8.22.

CNT filaments were found among the aerosol particles collected when applying high injector temperature during synthesis as seen in Figure 8.24c, e, and 8.24f. The morphology of particles produced at the different injector temperatures is shown in Figure 8.24c and d. A multitude of CNT filaments attached to clusters of nanoparticles was collected. The filaments were a few nanometers in diameter, as seen in Figure 8.24e and f, and are from a few nanometers to several micrometers in length.

Elemental analysis was performed on typical clusters as shown in Figure 8.24f; light and dark gray individual nanoparticles (marked by arrows) were collected at both high and low injector temperature; an example is indicated by the circle. The results of the elemental analysis are shown in Figure 8.25. While the Cu peaks and some contribution to the C and O peaks are expected to originate from the TEM grid, the Fe peaks are due to the collected particles and are likely from the dark gray particles that seem to include the Fe catalyst. This indicates that the particles are clusters of both carbon particles and iron catalyst particles coated with carbon.

Samples of MWCNTs synthesized at the low injector temperature were collected by scraping MWCNTs from the substrate. Grids were gently contacted with the MWCNTs and the collected MWCNTs were analyzed by TEM; the collected nanotubes are about 50 nm in diameter as can be seen in images of Figure 8.26a and b. In Figure 8.26b, the multiple layers inside a MWCNT can be seen at a fracture of a nanotube.

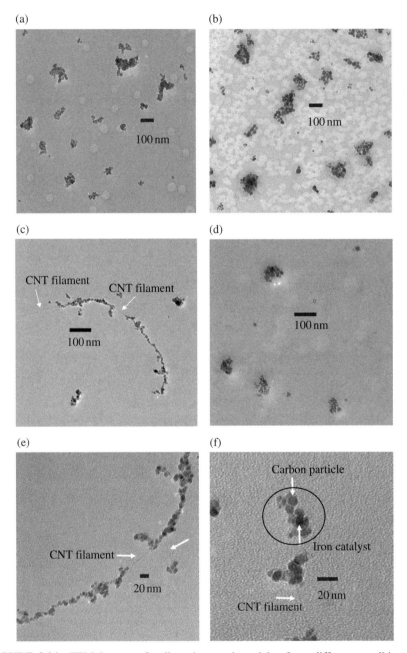

FIGURE 8.24 TEM images of collected aerosol particles from different conditions of MWCNT production. (a) L temp-A, Direct Mag:15600×; (b) L temp-B, Direct Mag:15600×; (c) H temp-B, Direct Mag:26000×; (d) L temp-B, Direct Mag:26000×; and (e, f) TEM images of MWCNT filaments collected from condition of high injector temperature and no substrate use (H temp-B). Reprinted with permission from Tsai et al. (2009b). Copyright 2009 *American Chemical Society*.

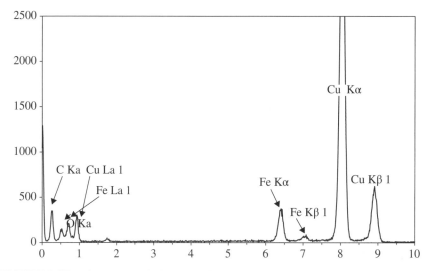

FIGURE 8.25 Elemental analysis results for the particles in Figure 8.24f. Reprinted with permission from Tsai et al. (2009b). Copyright 2009 *American Chemical Society*.

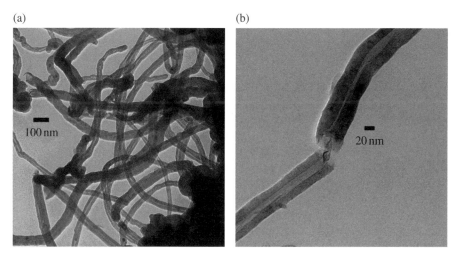

FIGURE 8.26 TEM images of MWCNT product produced from condition of low injector temperature and no substrate use (L temp-B). Reprinted with permission from Tsai et al. (2009b). Copyright 2009 *American Chemical Society*.

8.9.3.4 Particle Concentration Change in the Laboratory Background Particle concentrations in the laboratory background were measured before, during, and after synthesis of CNTs. The purpose was to evaluate the extent of particle release from the fume hood to the room. Typical measurements taken when using the low injector temperature for synthesis are shown in Figure 8.27, which shows the actual number concentration in the room. The concentration changes before furnace operation

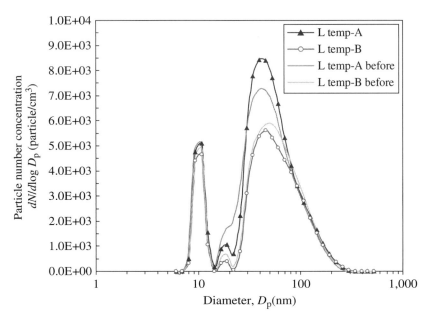

FIGURE 8.27   Particle number concentration at laboratory background. Reprinted with permission from Tsai et al. (2009b). Copyright 2009 *American Chemical Society.*

(no symbol line) and during operation (symbol line) varied by particle size. The consistent particle distribution pattern indicates that the concentration changes were due to the variability of the general ventilation system in the laboratory and not associated with nanoparticles released from furnace.

It should not be assumed that the excellent protection afforded by the fume hood used here would be provided by every chemical fume hood. The case study discussed in Section 8.8 found significant nanoparticle release associated with handling nanopowder when using a traditional conventional fume hood. However, detectable quantities of nanoparticles were not released into the room when using this particular fume hood during the furnace operation. This is likely due to certain favorable factors of the design and operation. The furnace was operated in a large (1.2 m (48 in.) wide × 0.76 m (30 in.) high) constant velocity fume hood; the sash was set at a 40 cm (16 in.) height location (0.7 m/s (140 ft/min) face velocity) during the whole operating process, and no worker was present at the hood face. More discussion of these design factors is presented in Section 9.1. We can conclude that a wider hood, intermediate face velocity and no worker standing in front of the hood can minimize the generation of turbulent airflow at the hood face, which could explain the good performance when using this hood.

8.9.3.5   We Learned that the Particle Release Varied by the Operating Conditions; However, What Would Be the Mechanism Causing the Formation of the Different Nanoparticles?   The injector temperature had a substantial effect on the particle size distribution; this is likely due to the dispersion of the catalyst solution, since at

high temperatures the evaporation of the catalyst is faster. The subsequent formation at higher temperatures of small clusters from an environment of gasified precursor will result in fine filaments which can travel through the reactor without depositing on its walls. When the liquid catalyst is introduced at low temperature, the droplets will not evaporate completely but form large particles. These particles cannot act as initiators for nanotube growth because of their size and they will only be coated with carbon instead of precipitating a nanotube.

However, when deposited on the reactor walls the catalyst droplets can decrease their diameter by evaporation and initiate substrate-bound growth. In other words, MWCNTs were well formed and attached on the substrate under low injector temperature operation. High injector temperature reduced the growth of MWCNTs on the substrate; instead, filaments were formed in the air and were released in the furnace exhaust. Thus, a low injection temperature has the double advantage of increasing MWCNT production while decreasing environmental release of nanoparticles as can be seen by the particle concentration shown in Figure 8.22.

8.9.3.6   What Do We Conclude from this Study?   The use of a substrate for MWCNT production results in fewer nanoparticles released into the air because when the CNTs are formed properly on the substrate that reduces the number of CNT filaments to be initiated in the air and then deposit on the wall or release in the exhaust. Consequently, a higher yield production for growing MWCNTs on a substrate could reduce the formation of aerosol filaments. However, the magnitude of nanoparticle release during all production conditions tested was very high with particle concentration usually exceeding 10^6 particles/cm^3 at the exhaust (source). While the toxicity of CNTs has received considerable attention, as discussed in Section 5.4, the toxicity of CNTs containing an iron catalyst, as found to be released in this study, still is not well understood.

The study demonstrated that the formation of filaments could be minimized while enhancing production by optimizing process conditions, but carbon particles were still formed in high quantity. This could cause adverse impacts on human health and the environment if the reactor exhaust is released directly into the outside air. For the normal operation of this process, the exhaust tube was connected through a water bath to prevent reverse airflow into the reactor; this had the added advantage that some aerosol particles in the exhaust air were captured in the water. The capture efficiency of the water bath is not known, however, and is likely to be fairly low, so that it would be desirable to filter the exhaust air. In conclusion, the operation of a CNT furnace in a well-designed ventilation hood or other ventilated enclosure is essential to prevent exposure of researchers to nanoparticles during CNT synthesis.

8.9.4   The Challenge and Brainstorming

Mario has been working with the CNT furnace placed in a fume hood for few years in a research lab. Recently, the university EHS officer examined the hood and the activities performed in the hood. Mario received a report notifying him that the hood was working well, in compliance with the requirements of the constant velocity hood, and no elevated aerosol particle concentration was detected outside the hood. However, after the inspection occurred, the furnace production was scaled up to synthesize a much larger quantity of MWCNTs. Mario noticed some dark spots on the surface of hood, the doorsill, the internal working surfaces, the furnace, and his gloves. He tried using a dry paper towel to clean up the spots, but this didn't seem to be effective; he then tried to wet the paper towel with water and to wipe dark spots, and he found that the wet paper towel did get black after wiping. However, he saw that the contaminated surface was still dark and appeared to be painted gray-black; he tried to wash the surface using more water, and the dark spots remained.

The dark spots kept appearing on the work surfaces and now the entire area looked dirty to Mario. He wondered if those black contaminants might be MWCNTs, or something else? He told himself that most of the MWCNTs should be attached to the substrate, so they are not likely to be airborne outside the furnace. However, he was worried that, no matter what comprised black material, there may be a problem for him to be working with tools and a work station contaminated with an unknown material.

8.9.5 Study Questions

1. What do you think Mario needs to do in this situation to help to reduce contamination and his worry?
2. Why cannot Mario wash clean the contaminated dark surface?
3. Do you think is there any chance that Mario carried contaminants back to his home? How likely is it to happen and in what way?
4. If the EHS officer was informed about this, what action should he/she take?

8.10   CASE STUDY: EXPOSURE FROM TWIN SCREW EXTRUSION COMPOUNDING

Polymer nanocomposites, which contain nanoparticles dispersed in a polymer matrix, provide improved properties at low filler loadings. These materials are already produced commercially, with twin screw extrusion being the preferred process for compounding the nanoparticles and polymer melts. Since commercial compounding (mixing) of nanocomposites is typically achieved by feeding the nanoparticles and polymer into a TSE, the airborne particles associated with nanoparticle reinforcing agents are of particular concern, as they can readily enter the body through inhalation. Recently, concerns have been expressed that airborne nanoparticles released during compounding might present significant exposure to extruder operators. To assess the impact of the nanoparticles during twin screw compounding of nanocomposites, researchers with experience in occupational and environmental health and polymer manufacturing monitored the compounding process for a model nanoalumina-containing nanocomposite using a FMPS.

Case theme: Compounding composites using nanometer-sized fillers and a polymer.

Operation: Compounding composites using an industrial-scale extruder with its typical configuration.

Investigation: Emission of particles, including their morphology and concentration magnitude from loading/feeding materials and compounding composites.

8.10.1   Materials and Production Process

Model nanocomposite systems consisting of nanoaluminum oxide (nanoalumina) (Al_2O_3) and polymethylmethacrylate (PMMA) or polyacrylonitrile-butadiene-styrene (ABS) were employed in this study. The nanoalumina particles were roughly spherical in shape with an average primary particle size ranging from 27 to 56 nm and a reported density of $3600 kg/m^3$. Each trial used 2.3 kg of polymer pellets and 160 g of nanoalumina particles.

Standard industrial equipment including a 30-mm corotating twin screw extruder (Werner & Pfleiderer, Model: ZSK-30) with a strand die, a single screw volumetric feeder, a twin screw volumetric feeder, a water bath, and a belt puller were used to compound polymer nanocomposites. During the twin screw extrusion process, the polymer pellets and nanoalumina particles were fed into the twin screw extruder where the polymer is melted and then mixed with the filler. The mixed melt was then forced through the die, forming a strand. This strand was then cooled and solidified as it was pulled through the water bath by the belt puller. Strand pelletizing equipment, which is typically attached to the line, was not used during most of the monitored trials (the strands were pelletized in a separate step).

As shown in Figure 8.28, the twin-screw extruder consists of two co-rotating screws in a metal barrel which contains heating elements and water cooling. This extruder has three feed or vent ports. Typically, polymer particles are fed from a single screw volumetric feeder into the feed port nearest the drive end of the screw. The polymer is melted by shearing elements in the first sections of the screws, allowing filler fed from a twin screw volumetric feeder into the second (middle) feed port to contact molten polymer. The polymer and filler is mixed between the second and third port, with the third port being available for venting of volatiles. Due to the incorporation of reverse pitched conveying elements in the screw program, however, most of the air fed into the extruder during feeding must be vented back through its feed port.

Consequently, three feeding methods were investigated in this study. For the "primary" feeding method, the polymer and nanoalumina were fed into the first port using a single screw volumetric feeder for the polymer and a twin screw volumetric feeder (with a stirrer) for the nanoalumina. The "secondary" feeding method was feeding of the polymer pellets using the single screw volumetric feeder into the first feed port and feeding of nanoalumina using the twin screw volumetric feeder into the second port. The third feeding method was feeding of premixed polymer pellets and nanoalumina using the single screw volumetric feeder into the first feed port.

A feeding tunnel was connected to the feed throat to smoothly load nanoparticles. As illustrated in Figure 8.28 (Tsai et al., 2008b), a local exhaust system had

(a)   (b)

FIGURE 8.28   (a) Layout of twin screw extruder and measurement locations. (b) Covering area at feeding port for engineering control. Reprinted with permission from Tsai et al. (2008b). Copyright 2008 *World Scientific*.

been installed on the TSE to help control contaminants given off by the process. It consists of a 30 cm (12 in.) diameter round hood connected to a flexible duct, with an exhaust air flow of 100 m³/h (60 ft³/min). During these experiments, the hood was positioned 30 cm (12 in.) above the extruder to collect polymer fumes given off from the melted mixture of polymer and nanoalumina; it was not placed near the feeding port.

8.10.2   Measurements

FMPS measurements were taken at background locations, source locations, and operators' BZs; in parallel to the FMPS real-time measurement, airborne nanoparticles were collected using polycarbonate filters fitted with filmed TEM grids driven by a personal air sampling pump. Measurement locations included a background location, a source location, and researchers' BZ (Fig. 8.28a). Background concentration measured before feeding nanoalumina into the twin screw extruder was used as the baseline for subtraction from the source concentration.

The nanoparticle aerosol filter sampler described in Section 7.2.4 and shown in Figure 7.16 was used in these experiments. Air flow was driven by a pump at a rate of 1.5–2.5 L/min, and particles were collected on the grid for analysis. Filter samples were analyzed for particle morphology and elemental composition.

8.10.3   Results

8.10.3.1 Airborne Nanoparticle Concentrations   Figure 8.29 shows that the particle number concentrations measured at the source location by three different feeding methods showed an increase in the range 50,000–150,000 particles/cm³ peaking at a particle size of 200 nm. For particles less than 50 nm, the maximum increase in particle number concentration was about 20,000 particles/cm³. The background concentrations obtained when feeding only polymer pellets were

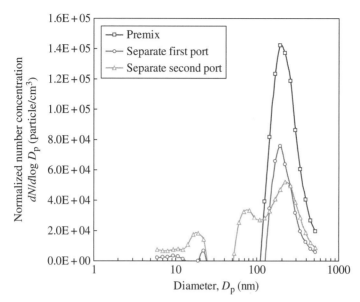

FIGURE 8.29 Increase in normalized number concentration at the source during feeding nanoalumina using different feed methods. Background aerosol associated with the polymer fumes has been subtracted from the distribution. Reprinted with permission from Tsai et al. (2008b). Copyright 2008 *World Scientific*.

subtracted to remove the independent concentrations caused by polymer fume. These airborne particle concentrations represent a significant increase in airborne nanoparticle concentration caused by the addition of nanoalumina compared to feeding only the polymer.

Separate feeding of nanoalumina to the first and second ports through the twin-screw feeder produced bimodal peaks of concentration increase at 5–100 nm. These small-particle peaks were not observed with premix feeding. Premix feeding, however, was a more energetic method compared to separate feeding, causing an increased release of larger aggregate nanoalumina particles into the room due to the dumping and transportation of a mixture of polymer pellets and nanoalumina particles. The increased particle concentration peaking at 200 nm with premix feeding was due to the release of aggregated nanoalumina particles.

These results suggest that the twin screws in the volumetric feeder were assisting in breaking up the nanoalumina agglomerates and nanoalumina particles were better dispersed in the polymer. TEM imaging of the resultant nanocomposites confirmed that feeding the nanoalumina through the twin screw feeder produced a larger percentage of primary particles and better dispersion of the nanoalumina (Ashter et al., 2009). This was the desired result, but more individual nanoalumina particles were released into the air. Nevertheless, the magnitude of particle release through separate feeding to the first port was less energetic and presented the lowest exposure among these three feeding methods.

8.10.3.2 Particle Morphology and Elemental Analysis Particles collected in parallel with FMPS measurements while feeding nanoalumina are shown in Figure 8.30. Collected particles on polycarbonate filters and TiO_2-coated TEM grids were analyzed by SEM and STEM, respectively. The SEM and STEM results in conjunction with EDS analysis were consistent with the FMPS data (shown in Fig. 8.29) with respect to the fraction of the total particles above background that were identified as alumina. Many aggregated nanoalumina particles sampled at the source were around 200 nm in diameter as seen on the SEM and STEM images. Nanoalumina particles at the workers' BZ were collected and are shown in Figure 8.30d and e. In Figure 8.30f, a mixture of particles collected on TEM grid at the source can be seen on the image. The light gray particles are carbon aerosols coming from the mixture of background airborne particles and the nanoalumina particles are indicated by arrow marks.

The size of nanoalumina particles found on the filter samples is in the range from nanometers to several micrometers as shown in Figure 8.30a–e. This could indicate that micrometer-sized agglomerated nanoalumina particles exist in the air while the primary size of nanoalumina particles in the bulk material is less than 100 nm. Using just an instrument such as the FMPS that only measures particles in the nanometer size range would not have detected the agglomerated particles that were formed and present in the air. This leads to the important conclusion that the measurement of particle concentration and size distribution from nanometer to micrometers is important to fully investigate the behavior of aerosols formed from processing nanoparticles. In addition, information on particle morphology and elemental composition frequently is important to completely characterize an aerosol formed by processing nanoparticles.

8.10.3.3 The Status of Controls The feeding port area of the extruder was operated using a twin screw feeder feeding nanoalumina particles into an open top feeding throat of a single screw feeder feeding polymer pellets from the hopper as shown in Figure 8.29b. Nanoalumina particles were loaded through the feeding throat about 18 cm (8 in.) above the receiving port of extruder into the mixer of polymer as shown in Figure 8.28a. Isolation or a cover was not available for this feeding area where the highest nanoalumina particle concentration was detected.

The existing local exhaust hood was poorly designed and inadequate to collect either the polymer fumes or the nanoalumina particles given off by the process. The measured air flow through the exterior hood ($100 m^3/h$) was much too low, resulting in a capture velocity at the equipment surface of 3 cm/s (6 ft/min) as calculated using DallaValle's equation (Burgess et al., 2004). This velocity is lower than the random air motion present in the vicinity of the equipment, so that there would be little or no particle movement toward the hood. The poor performance of the ventilation system is consistent with the very high airborne particle concentrations measured at both worker's BZ and the source location in these extrusion trials.

8.10.3.4 A Test of Applying Controls The magnitude of nanoparticle exposure associated with applicable controls was measured before and after applying engineering and administrative controls to the compounding process. Engineering control was evaluated by completely covering feeding outlet of the twin screw feeder and the feeding throat of

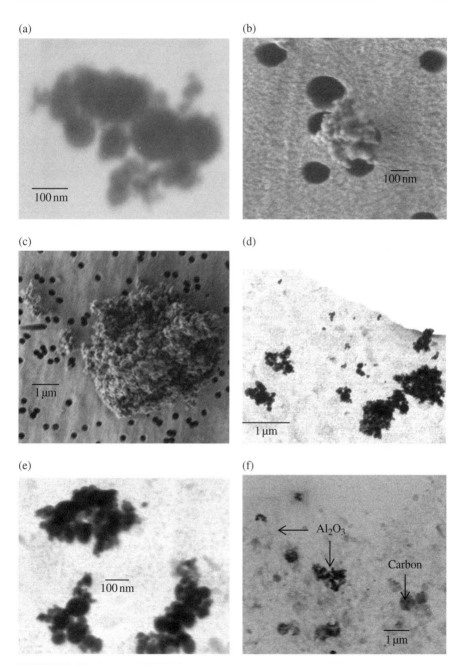

FIGURE 8.30 SEM and STEM images of samples taken in parallel with FMPS measurements at the source and the breathing zone locations: (a), (b), (c), and (f) at source; (d) and (e) at breathing zone; (b) and (c) are SEM images of polycarbonate filters, others are STEM images of filmed TEM grids. Reprinted with permission from Tsai et al. (2008b). Copyright 2008 *World Scientific*.

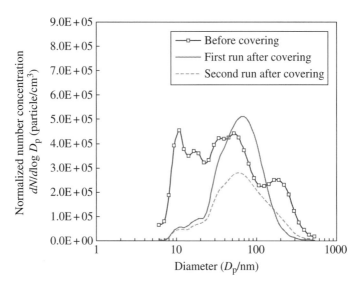

FIGURE 8.31   Normalized number concentration at the source before and after covering feeding port. Reprinted with permission from Tsai et al. (2008b). Copyright 2008 *World Scientific.*

single screw feeder as shown as circle mark in Figure 8.28b. Administrative control was evaluated by housekeeping, that is, using water-based cleaning to thoroughly wash the floor and remove dust on equipment throughout the laboratory.

It was not possible to improve the ventilation system for this case. In order to investigate possible controls other than ventilation to reduce particle release from the feeding area, aluminum foil was used to fully cover the open top of the feeding throat and the open mouth of the twin screw nanoalumina particle feeder (Fig. 8.28b). Two measurements were taken at successive times at the source location after covering the feeding area. As shown in Figure 8.31, particles released in size ranges of less than 30 nm and above 200 nm were dramatically reduced at the first measurement compared to the measurements prior to covering the feeding area. The second measurement at the source location taken 30 min later showed a continuing significant reduction of particle concentration in the size range 30–100 nm.

The primary solution to reduce exposure to nanoalumina and polymer fume nanoparticles is to isolate the releasing source. In order to eliminate particle release at the source, sealing and covering the open feeding throat and gaps can efficiently block the route for releasing nanoparticles and results in a decrease of nanoparticle concentrations in the laboratory. In other words, an isolated and enclosed nanoparticle feeding system is the primary solution to efficiently control particle release to low levels.

8.10.4   The Challenge and Brainstorming

In reading the manufacturer's information about the alumina nanoparticles being used in his compounding process, Nick found that nanoalumina is very hydrophilic. Nick was cheered by this news and quickly stepped into the compounding room with a sprayer bottle. Nick's colleague stopped him and questioned what was in his bottle. "Water," Nick responded. Soon he had sprayed water all around the compounding machine where he had observed a considerable amount of white nanoalumina on the equipment surface. Nick and his colleague did not observe much difference, except that some surfaces felt somewhat damp. Nick said happily to his colleague, "I guess those nanoalumina aerosols get collected by the water droplets. We should have less nanoalumina in the air."

Later in the week, Nick's colleague came to him and complained that the composites they made few days ago had degraded mechanical properties and his analysis showed the fillers (nanoalumina) were not well distributed within polymer. Nick responded: is the water sprayed on that day relevant to this result?

8.10.5   Study Questions

1. What scientific reason did Nick come up with to suggest spraying water around the compounding equipment?
2. Can spraying water help to reduce nanoalumina aerosol exposure? How?
3. Is the water spray relevant to the distribution of fillers in polymer? Is this in conflict to Nick's goal to reduce aerosol nanoalumina?
4. What other controls can Nick consider to reduce his exposure?

APPENDIX 8.A.1   Normalized Size Distributions

An aerosol spectroscope measures some property of an aerosol (number, surface area, etc.) as a function of the particle diameter; here, we will use the FMPS as an illustrative example. The FMPS measures particle number concentration as a function of particle diameter over the range 5.6–560 nm, in 32 size channels. To do this, the instrument manufacturer divides the measured range of particle diameters into 32 size channels and counts the number of particles in each channel. Since the range of particle diameters covers two orders of magnitude, there are 16 size channels per decade of diameter.

The simplest procedure would be to divide the size range from 5.6 to 560 nm into 32 channels of equal width, but that is not done. The reason is that aerosols typically follow a log-normal size distribution, so that fewer and fewer particles are found as the diameter increases. Therefore, the larger size channels will have relatively fewer particles. To compensate for this, the instrument manufacturer divides the size range into channels as a function of the *log* of the diameter; this makes the width of the channels increase as a function of the diameter, so that relatively equal numbers of particles are found in each size channel.

In our specific example, the log of 5.6 is 0.75 and the log of 560 is 2.75, so the range of log d from 0.75 to 2.75 is divided into 32 equal log ranges and then converted back to diameter to give the instrument's size channels. In this case, the width of each size channel increases by about 15.5% (i.e., $1.155^{32} = 100$, the ratio of the largest to the smallest diameter).

The logarithmic distribution of the size channels affects the way the measured data must be displayed. Since the channels become wider as the diameter increase, plotting of the raw data would not accurately reflect the actual aerosol size distribution. To account for this, the count data are *normalized* by dividing the number of particles counted by the *width* of the channel; the resulting number is the number of particles per nanometer of diameter. Since the width of the size channels is increasing logarithmically, this is equivalent to dividing the number of particles counted by the log of the diameter. Thus, in all of the figures in this book that present particle size distribution data, such as Figure 8.14 for example, the vertical axis is labeled "particle number concentration, $dN/d\log D_p$ (particles/cm^3)" which should be read as "particle number concentration, normalized for the width of the size channel."

REFERENCES

Ashter A, Tsai S, Lee JS, Ellenbecker M, Mead J, Barry C. Effects of nanoparticle feed location during nanocomposite compounding. Polymer Eng Sci J 2009;50:154–164.

Bottini M, Magrini A, Bottini N, Bergamaschi A. Nanotubes and fullerenes: an overview of the possible environmental and biological impact of bio-nanotechnologies. Med Lav 2003;94:497–505.

Burgess WA, Ellenbecker MJ, Treitman RD. *Ventilation for Control of the Work Environment.* New York: Wiley-Interscience; 2004.

Collins PG, Avouris P. Nanotubes for electronics. Sci Am 2000;283:67–69.

Ferreira T, Rasband W (2012) Image J user guide. National Institutes of Health Version IJ 1.46r.

Guo T, Nikolaev P, Thess A, Colbert D, Smalley R. Catalytic growth of single-walled nanotubes by laser vaporization. Chem Phys Lett 1995;243:49–54.

Hinds WC. *Aerosol Technology—Properties, Behavior, and Measurement of Airborne Particles.* New York: Wiley-Interscience; 1999.

Iijima S. Helical microtubules of graphitic carbon. Nature 1991;354:56–58.

Kumar P, Fennell P, Britter R. Measurements of particles in the 5-1000 nm range close to road level in an urban street canyon. Sci Total Environ 2008a;390:437–447.

Kumar P, Fennell P, Langley D, Britter R. Pseudo-simultaneous measurements for the vertical variation of coarse, fine and ultrafine particles in an urban street canyon. Atmos Environ 2008b;42:4304–4319.

Kumar P, Fennell P, Symonds J, Britter R. Treatment of losses of ultrafine aerosol particles in long sampling tubes during ambient measurements. Atmos Environ 2008c;42:8819–8826.

Longley ID, Gallagher MW, Dorsey JR, Flynn M. A case-study of fine particle concentrations and fluxes measured in a busy street canyon in Manchester, UK. Atmos Environ 2004;38:3595–3603.

McCarrie KM, Winter R (2003) Properties of epoxy-clay nanocomposite adhesives for bonded strap joints. San Jose (CA): San Jose State University. San Jose State Report.

Mirero, Inc. GAIA Blue. Available at http://www.gaia-zone.com/pro_gaiaBlue.htm. Accessed January 3, 2013.

Motulsky H. *Intuitive Biostatistics—A Nonmathematical Guide to Statistical Thinking.* New York: Oxford University Press; 1995.

Popendorf W. *Industrial Hygiene Control of Airborne Chemical Hazards*. Boca Raton, FL: Taylor & Francis; 2006. p 1.

Rappaport SM, Kupper LL. *Quantitative Exposure Assessment*. El Cerrito, CA: Stephen Rappaport; 2008.

Tsai SJ, Ashter A, Ada E, Mead J, Barry C, Ellenbecker MJ. Airborne nanoparticle release associated with the compounding of nanocomposites using nanoalumina as fillers. Aerosol Air Qual Res 2008a;8:160–177.

Tsai SJ, Ashter A, Ada E, Mead J, Barry C, Ellenbecker MJ. Control of airborne nanoparticle release during compounding of polymer nanocomposites. NANO 2008b;3:1–9.

Tsai SJ, Ada E, Isaacs J, Ellenbecker MJ. Airborne nanoparticle exposures associated with the manual handling of nanoalumina and nanosilver in fume hoods. J Nanopart Res 2009a;11:147–161.

Tsai SJ, Hofmann M, Hallock M, Ada E, Kong J, Ellenbecker MJ. Characterization and evaluation of nanoparticle release during the synthesis of single-walled and multi-walled carbon nanotubes by chemical vapor deposition. Environ Sci Technol 2009b;43:6017–6023.

Wang J, Flagan RC, Seinfeld JH. Diffusional losses in particle sampling systems containing bends and elbows. J Aerosol Sci 2002;33:843–857.

Xiang R, Luo G, Qian W, Wang Y, Wei F, Li Q. Large area growth of aligned CNT arrays on spheres: towards the large scale and continuous production. Chem Vap Dep 2007;13:533–536.

9

CONTROL OF OCCUPATIONAL EXPOSURES TO ENGINEERED NANOPARTICLES

As discussed in the Chapter 8, the evaluation of exposures to engineered nanoparticles (ENPs) is quite difficult at this time, with a great deal of uncertainty as to the optimum methods to be used. The *control* of such exposures, on the other hand, is more straightforward. The classical methods to control occupational exposures to airborne and dermal contaminants, developed over the last century, can be applied to ENPs. That being said, it is extremely important to recognize that the *level* of control typically attained for other contaminants may not be adequate for ENPs. As an example, the performance of an exhaust hood in capturing aerosols, as measured by *mass* concentration, may be considered satisfactory but that same hood might allow a significant *number* of ENPs to escape. This chapter will describe the classical methods used to control exposure to particles, with an emphasis on the aspects of their performance that are crucial to ensure satisfactory performance against ENPs. It will first cover engineering controls for airborne particles, followed by a discussion of controls specific to dermal exposure, and conclude with a discussion of administrative controls and good practices applicable to both exposure routes.

9.1 CONTROL OF AIRBORNE EXPOSURES

9.1.1 General

Given the current uncertainty about the toxicity of ENPs, it is very important that exposures be kept to the lowest practical level. It is important that the full range of

Exposure Assessment and Safety Considerations for Working with Engineered Nanoparticles,
First Edition. Michael J. Ellenbecker and Candace Su-Jung Tsai.
© 2015 John Wiley & Sons, Inc. Published 2015 by John Wiley & Sons, Inc.

occupational hygiene controls be utilized to limit exposures. Following the hierarchy of controls, engineering controls should form the first line of defense. Administrative controls are also extremely important and should always be used in conjunction with engineering controls.

Engineering controls include any physical change to the process or workplace that reduces contaminant emissions and subsequent worker exposure. The primary engineering controls are briefly described in the following texts in order of preference.

9.1.1.1 *Process Change/Chemical Substitution*   This is the first control to consider, since changing a process to reduce emissions or substituting a nonhazardous or less-hazardous chemical or material for a more-hazardous one is a *permanent* solution to the exposure problem. Process change/chemical substitution has the added great advantage of eliminating both the worker exposure and emissions to the general environment. For example, if a reaction is performed using a chlorinated solvent such as trichloroethylene (TCE), a laboratory fume hood may be needed to reduce worker exposure (see Section 9.3.2). While this may solve the worker exposure problem, the TCE vapors in the hood exhaust air become an air pollution problem unless the exhaust air is cleaned before discharge (not likely for most laboratory fume hoods). If the reaction conditions are changed, however, so that water can be used as the solvent, this chemical substitution eliminates the need for using the fume hood and is a permanent solution to both the worker exposure and the environmental release.

For workers performing research with nanomaterials, it is probably not feasible to substitute another material for the nanomaterial itself, since that is the focus of the research. It may be possible, however, to change some aspects of the *process* in a way that reduces nanoparticle release. For example, working with nanoparticles suspended in a liquid is a significant improvement over working with them in dry powder form. Another example recently occurred in our work with a laboratory growing carbon nanotubes (CNTs) by chemical vapor deposition, as discussed in the case study of Section 8.9. We found that by optimizing the furnace reaction temperature, the production of CNT was maximized, while at the same time minimizing the release of CNTs in the furnace exhaust (Tsai et al., 2009a).

Process change as a control strategy has been the focus of much research in recent years. It goes by many names, including pollution prevention, toxics use reduction, cleaner production, and so on. When the focus is chemical synthesis, the term most commonly used is green chemistry. The University of Oregon has an active research program in green nanotechnology; their recent report (Matus et al., 2011), written together with the American Chemical Society's Green Chemistry Institute, describes research opportunities and challenges in this important field.

An important disincentive to using process change as a control strategy is the fact that such changes must be individually designed for each industry, and indeed each process within that industry. For this reason, case studies are an important source of information on process changes that have been found to be effective in specific situations, including nanotechnology (Ellenbecker and Tsai, 2011). The US Environmental Protection Agency's Office of Pollution Prevention and Toxics and several state pollution prevention programs, such as the Institute of the Environment and Sustainability at UCLA, the New York State Pollution Prevention Institute at the Rochester Institute of Technology, and the Massachusetts Toxics Use Reduction Institute at the University of Massachusetts Lowell, have published many such case studies. Readers are encouraged to search the web sites of these and similar agencies for information regarding good process change opportunities for the nanomaterial of interest to them.

9.1.1.2   Isolation   Isolation refers to the physical isolation of a process using hazardous materials from the rest of the workplace. A common example of isolation used in chemistry laboratories is the use of designated storage cabinets for flammables, acids, bases, etc. Isolation has important advantages of limiting the *scope* of a potential hazardous exposure to the smallest physical area and the fewest number of people. If exhaust ventilation is also used (see the following section), isolation has further advantage of limiting the area where ventilation is required, thus limiting the amount of air flow required. The modifications of the twin screw extruder discussed in the case study of Section 8.10 are a good example of the combined use of isolation and ventilation.

It is very important when working with potentially toxic engineered nanomaterials that the work to be performed in as isolated space as possible with as few people in that space as possible. This is particularly important in educational settings, where many students may have access to the laboratory areas. This type of isolation can also be considered an administrative control, as described in Section 9.3.

9.1.1.3   Ventilation   There are two distinct types of ventilation systems that can be used to control exposure to airborne contaminants, that is, local exhaust ventilation (LEV) and general exhaust ventilation (GEV), also called dilution ventilation. In the hierarchy of controls, LEV is preferred over GEV. A GEV system typically uses a wall or ceiling fan to exhaust air from a room; any contaminants given off into the room air will mix with the fresh replacement air entering the room and will eventually leave the space with the exhaust air. The introduction of fresh air with GEV thus does not *eliminate* worker exposure to the contaminant, but the GEV does reduce the airborne concentration of the contaminant, thus *reducing* the exposure. Only general principles of exhaust system design are discussed here; readers desiring more detailed information should consult a ventilation design textbook (Burgess et al., 2004; Goodfellow and Bender, 1980).

A typical LEV system is shown in Figure 9.1. Here, an exhaust hood is placed next to or encloses the contaminant source; air flowing into the hood entrains the contaminants and carries them through the duct, where they are either removed by an air cleaner or vented to the atmosphere. The most common LEV system used in research laboratories is the laboratory fume hood (Fig. 9.2); this is an example of an *enclosing* hood, since the hood itself physically surrounds the contaminant source. The fume hood, with its moving sash, is actually only a partial enclosure; an example of a complete enclosure is the glove box (Fig. 9.3).The other basic hood type, the *exterior* hood, is placed adjacent to the contaminant source; hoods of this type are also commonly found in laboratories (Fig. 9.4).

In general, enclosing hoods are preferred to exterior hoods, since the contaminants are given off inside the hood itself and the hood provides a barrier between the worker and

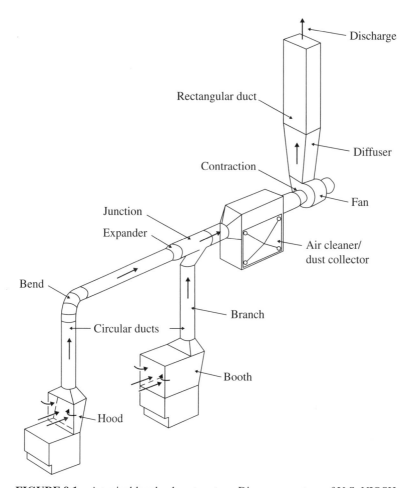

FIGURE 9.1   A typical local exhaust system. Diagram courtesy of U.S. NIOSH.

the contaminant. Sufficient air flow must be provided through any openings in the enclosure to ensure that the contaminants don't escape the hood. This is sometimes problematic for enclosing hoods with large openings, such as the fume hood—this is discussed
more fully in the following section. Exterior hoods are less preferred because they must
create a capture velocity at the point of contaminant generation to capture contaminant
and draw it into the hood and are thus subject to disruptive crossdrafts.

9.1.2   Laboratory Fume Hoods

9.1.2.1   General Considerations   Laboratory fume hoods are a commonly
used form of LEV, designed to protect researchers from chemical exposures
while experiments are performed. Fume hoods are ventilated enclosures located
at bench height and large enough so that simple chemical reactions or

FIGURE 9.2   A typical laboratory fume hood. Photo by S. Tsai.

FIGURE 9.3   A glove box. Photo by S. Tsai.

FIGURE 9.4   An exterior fume hood. Photo courtesy of U.S. NIOSH.

nanoparticle manipulations can be carried on inside; typical laboratory fume hoods are shown in Figure 9.5. There are many different fume hood designs, but the most common categories, as discussed in the following, are the constant flow hood, the bypass hood, and the constant velocity hood. All fume hoods have certain common design elements, including an exhaust fan to move air through the hood, a moving sash, exhaust slots, and a horizontal work surface. The sash can move in either a vertical or horizontal direction; the vertical sash design is much more common.

The exhaust air flow through the hood is meant to entrap any toxic molecules or aerosol particles given off from the process being performed inside the hood and carry them to the exhaust system outlet, thus preventing them from escaping the hood and entering the worker's breathing zone. Fume hoods are usually considered by laboratory workers to be the health and safety workhorse in their work environment; they are ubiquitous in laboratories, and workers commonly assume that, if they are working inside a fume hood, they will be adequately protected from the inhalation of toxic gases, vapors, and aerosols. As discussed below, this assumption may not always be warranted. Proper performance of a fume hood depends on the hood *design*, the hood *operation*, and the *work practices* used in the hood. If any of these

(a)

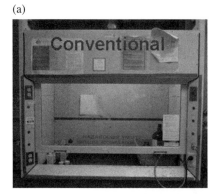

(b)

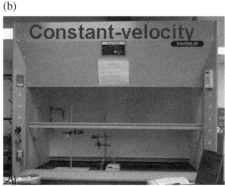

(c)

FIGURE 9.5   Common laboratory fume hood designs: (a) conventional (constant-flow) hood; (b) bypass hood; and (c) constant-velocity hood. Photos by S. Tsai.

factors are inadequate, significant contaminant release and subsequent worker exposure can occur; this is particularly true for nanoparticles.

A crucial design element for any fume hood is the *face velocity*, defined as the average velocity of the air moving through the face of the hood at the sash opening. The relationship between hood average face velocity and hood air flow is

$$Q = vA \qquad (9.1)$$

where Q = hood air flow (m³/s), v = average hood face velocity (m/s), and A = hood sash open area (m²).

The constant flow hood constitutes the oldest, simplest fume hood design. It is so-called because the exhaust fan introduces a constant air flow Q through the sash opening. For this hood design, the face velocity is lowest when the sash is wide open (A is at its maximum value); when the sash is lowered, the face velocity increases as given by Equation (9.1).

The constant velocity hood, also called a variable air volume (VAV) hood, uses a control system to detect the sash position and increase the air flow when the sash is raised and decrease it when the sash is lowered, thus maintaining a constant face velocity. The bypass hood uses a different approach to attempt to maintain a constant hood face velocity. This hood incorporates a bypass grille located above the sash opening, as shown in Figure 9.5c. When the sash is wide open, it blocks the bypass grille, allowing all of the air to flow through the hood opening. As the sash is lowered, it uncovers increasingly greater amounts of the bypass grille, allowing increasing amounts of air to flow through this alternative path. If it is designed properly, the amount of air flowing through the bypass grille is just sufficient to maintain a constant hood face velocity. Typically, however, this constant velocity can only be maintained over a certain part of the sash's total range. Another disadvantage of the bypass hood is that any air flow through the bypass grille is essentially wasted; since the air in laboratories typically is air-conditioned, this represents a significant waste of energy.

Saunders (1993) and DiBerardinis et al. (1992) describe basic elements that all fume hoods should incorporate. Before using any hood for nanopowder manipulation, it should *at a minimum* include the following design elements:

- A minimum width of 120 cm (4 ft) (wider is better);
- A minimum sash open height of 80 cm (30 in.);
- A bottom-front airfoil;
- Side walls that are smooth, rounded, and tapered toward the inside of the sash opening (Schulte et al., 1954);
- A sash that is easily movable over its entire range of motion;
- A sash that holds its position over its entire range of motion.

In addition, the following factors relative to the hood *location* are very important for proper hood performance:

- The hood should not be located next to any laboratory entry door or any other high-traffic location;
- The best hood location is in a corner of the laboratory, opposite the main laboratory entrance;
- The hood should be located at least 1.5 m (5 ft) from any heating, ventilating, and air-conditioning (HVAC) air supply grille; 3 m (10 ft) is preferred.

The following factors relative to *work practices* are also crucial to proper hood performance:

- The hood should only contain the equipment and materials necessary for the operation being performed—excessive equipment and bottles of chemicals will disrupt the normal hood airflow pattern and degrade hood performance.
- The hood sash should be kept wide open only during equipment setup; during actual use, the sash should be lowered to the position that gives proper hood face velocity, as discussed in the following.

- Equipment should be located at least 15 cm (6 in.) behind the sash opening (many hoods have a recessed floor starting at this distance, to encourage proper use).
- When working in the hood, the user should minimize arm movements and make all such movements in a slow, smooth manner. Traffic past the hood should be minimized when nanopowders are being manipulated; research has shown that the passage of a person past the hood face at walking speeds creates a turbulent wake sufficient to pull contaminants from the hood (Johnson and Fletcher, 1996). During experiments, when no worker access is required, the sash should either be kept in the same position as when work is performed (constant flow and bypass hoods) or be lowered to the fully closed position (constant velocity hoods).
- When working in the hood, users should wear appropriate gloves, clothing and personal protective equipment as discussed in Section 9.3.

9.1.2.2 Hood Face Velocity   All of the above factors can be easily ascertained and/ or controlled by the hood user. However, the single factor that is most important for proper hood performance—hood face velocity—is beyond the ability of the user to determine; this must be evaluated and controlled by the facility's health and safety staff. A great deal of research has been undertaken to understand the role that face velocity plays in proper fume hood performance; research specifically relative to handling nanopowders will be described in the following section. First we will present the current consensus on proper hood face velocity, as given by scientists (Altemose et al., 1998; Burgess et al., 2004; DiBerardinis et al., 1992; Saunders, 1993), government agencies (OSHA, 1992), and professional organizations (ACGIH, 2007; ANSI/AIHA, 2002). None of these references specifically discuss proper face velocities for nanoparticle control. Recently, the authors of this book and NIOSH published guidelines for nanomaterial research laboratories containing recommendations regarding fume hood face velocity (NIOSH, 2012); these recommendations are largely based on our research discussed in the following text (Tsai et al., 2009b, 2010).

It is the consensus of the sources referenced above that the average face velocity for a laboratory fume hood should be in the range 0.4–0.6 m/s (80–120 ft/min). Some experts recommend a narrower range; for example, DiBerardinis et al. recommend that a face velocity of at least 0.5 m/s (100 ft/min) be maintained at all times, with a face velocity of 0.6 m/s (120 ft/min) used for highly toxic materials. Older publications tended to recommend wider ranges of acceptable face velocities. Caplan and Knutson, as a result of their ground-breaking work in the 1970s to develop a quantitative method to evaluate fume hood performance (Caplan and Knutson, 1978), recommended using face velocities as low as 0.3 m/s (60 ft/min). At the other extreme, OSHA in its 1992 regulations recommended using fume hoods with face velocities greater than 0.8 m/s (150 ft/min) when handling any of the then-existing list of 13 OSHA carcinogens in a fume hood (OSHA, 1992).

The American Conference of Governmental Industrial Hygienists (ACGIH), in their widely used design book *Industrial Ventilation: A Manual of Recommended Practice*, has changed the recommended face velocity several times since the first edition was published in 1951. Early editions recommended that face velocities be between 0.5 and 0.8 m/s (100–150 ft/min), with the higher value used for higher toxicity materials (ACGIH, 1966). This recommended range continued to be published until the 16th edition in 1980 (ACGIH, 1980); in the 17th edition, published in 1982 (ACGIH, 1982), the recommended range was widened to 0.3–0.8 m/s (60–150 ft/min), presumably based on the research of Caplan and Knutson (1978). This range was recommended until the 22nd edition, published in 1995 (ACGIH, 1995), when the range was limited to 0.4–0.5 m/s (80–100 ft/min), the current recommendation.

Another fairly recent recommendation is made by Fletcher, who authored the relevant section in the *Industrial Ventilation Design Guidebook* (Goodfellow and Tahti, 2001). He reviewed many of the face velocity recommendations and concluded the range suggested by ACGIH would seem to be generally acceptable with caution being exercised at the lower end of the range. Since ACGIH recommends 0.4–0.5 m/s (80–100 ft/min), rather than 0.4–0.6 m/s (80–120 ft/min), and caution is needed at the lower end, Fletcher points toward 0.5 m/s (100 ft/min) as the most desirable fume hood average face velocity.

In addition to the *average* face velocity, it is important that the airflow be *distributed evenly* across the hood face. The consensus is that the face velocity at any point on the hood face be within ±20% of the average face velocity, although some sources recommend a stricter standard of ±10% (ACGIH, 2007).

It is important to understand why the current consensus recommendation for face velocity falls in such a narrow range. The large amount of research on this topic has shown that values below this range are subject to room air currents and rapid operator movements that can pull nanoparticles out of the hood, while values above this range create excessive turbulence within the hood itself and in the wake of the worker that can also pull nanoparticles out of the hood.

It is well established that air flowing past a human body creates downstream turbulent wakes (Flynn and Ljungqvist, 1995; Kim and Flynn, 1991). In the case of a fume hood, the turbulent wake eddies are created between the worker and the hood face by exhaust air drawn into the hood past the worker (Fig. 9.6). These eddies, also called a bluff-body wake (Tseng et al., 2006), can draw contaminants from eddy regions in the hood near the sash opening where nanoparticles can accumulate due to the interior hood eddy described below.

Poorly designed hoods, such as the conventional hood discussed in the case study, Section 8.8, can have large vertical eddies that can carry nanoparticles released by the handling tasks up the back of the hood , across the top, and then down toward the sash opening, where they can interact with room air currents and/or turbulent wake eddies (Fig. 9.7). A recent study showed that complicated turbulent flow patterns were induced around the bottom edge of the sash, side rails, and doorsill, where the containment leakage was highest in a conventional fume hood (Tseng et al., 2006). When considering both the exterior eddy, which is a strong function of the hood face velocity, and the interior eddy, which depends on hood design and the position of the sash, it is apparent that particle release must be a complicated function of the interaction between the hood vertical eddy, the worker wake eddy, external crossdrafts, the position of the sash, hood face velocity, and worker activities, such as hand and arm motions, that may be occurring at the interface between the two eddies.

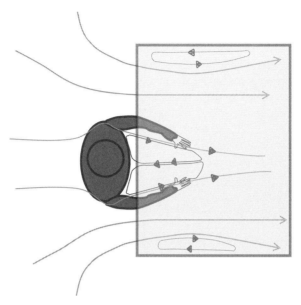

FIGURE 9.6   Wake region and air flow pattern. Drawing produced by W. Heitbrink, U.S. NIOSH, based on research by Kim and Flynn (Flynn and Ljungqvist, 1995; Kim and Flynn, 1991).

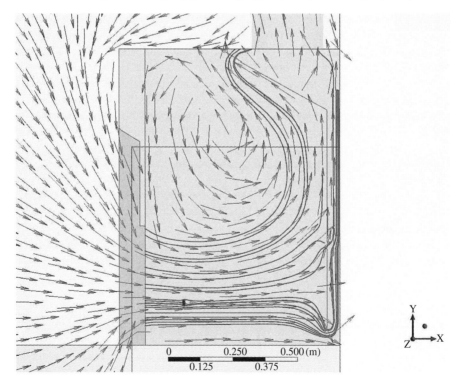

FIGURE 9.7   Vortex region of traditional fume hood. Computational fluid dynamics (CFD) model courtesy of K. Dunn, U.S. NIOSH.

Tests that compared the performance of different hood designs, which are discussed in Section 9.5, showed that the conventional fume hood (Fig. 9.5a) had the highest particle release to the breathing zone among the three hood types. A large vortex region circulating downward and reaching the sash bottom as illustrated in Figure 9.7 was observed in the conventional hood using a fog test; the vortex carried many particles into the wake region which then flowed out of the hood and toward the worker's breathing zone (Pathanjali and Rahman, 1996).

9.1.2.3 *Working with Nanopowders in Fume Hoods*

It is commonly thought that manipulating solid materials inside a chemical fume hood that meets the above guidelines offers the researcher good protection. However, extensive research performed in our laboratory (Tsai et al., 2009b, 2010), which is described in detail in Section 8.8, has demonstrated that this definitely is not the case for NP powders, where releases that are nondetectable on a mass basis can have a very high number concentration. Experiments performed on constant volume and bypass hoods have demonstrated that working with the sash either too low or too high can cause NPs released inside the hood to escape from the hood and cause exposure to the researcher.

When the sash is too high, the face velocity can fall below the recommended minimum of 0.4 m/s (80 ft/min); this low face velocity and the large opening created by the high sash allow random air currents to enter the hood, entrain airborne NPs, and carry them out of the hood. When the sash is too low, the face velocity can exceed the maximum recommended value of 0.6 m/s (120 ft/min); this causes a strong turbulent wake in the space between the worker and the hood face, which can pull airborne NPs from the hood. An example of such turbulence is shown in Figure 9.8, taken using a constant-flow hood with a mannequin, a laser sheet light source and a smoke machine. Figure 9.8a uses a horizontal light sheet; the turbulence is clearly visible between the mannequin's arms, the airstreams circulate toward the middle of arms from both arm sides. Figure 9.8b has the laser oriented vertically and shows a circulatory airflow pattern toward the middle of two arms from top and then outward under each arm so that air continuously circulates around each arm.

(a)   (b)

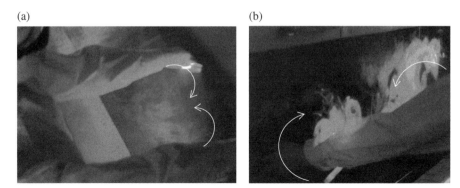

FIGURE 9.8 Turbulence patterns between a mannequin and a laboratory fume hood: (a) horizontal plane and (b) vertical plane. Photos by S. Tsai.

As discussed earlier, the effectiveness of a laboratory fume hood in protecting the user from chemical or nanoparticle exposure is dependent on the proper placement of the hood sash and the resulting face velocity into the hood. Every research laboratory should have a procedure for routinely monitoring laboratory fume hood performance and determining proper hood sash height to give good performance. Each fume hood should be clearly marked with the proper hood sash location; depending on the hood design, this could be a single location or a range of locations. It is extremely important that processes be carried out inside the hood with the sash at the single position or within the range of positions specified by the laboratory environmental health and safety (EHS) office. If process requirements require that these conditions cannot be met, researchers should contact their health and safety office for an evaluation of potential exposures.

9.1.2.4 New Fume Hood Designs   Researchers are always designing new fume hoods which may offer improved performance when handling NP powders. Considerable research has recently been dedicated to fume hoods that operate at lower face velocity, in order to effect energy savings. Examples include the low-flow hoods manufactured by LabCrafters (Taylor, 2004) and the Berkeley hood (Tschudi et al., 2004). There are no reports in the literature on the effectiveness of low-flow fume hoods when handling NP powders. A completely new hood design approach is taken by the air-curtain hood (Huang et al., 2007), which uses a downward air jet emanating from a double-pane sash to isolate the interior of the hood from the exterior environment. Recent work in our laboratory indicates that this hood is extremely effective at containing airborne NPs (Tsai et al., 2010).

9.1.2.5 Fume Hood Recommendations   When working with dry nanopowders, it is important to work in a ventilated enclosure of some sort. Research indicates that the level of protection afforded will vary greatly depending on the ventilation system used and the manner in which it is used. Alternative ventilation approaches, as described in Section 9.1.3, may offer better protection than a laboratory fume hood. If using a fume hood, it is important to use the *best* fume hood available. Summarizing the aforementioned discussion, a constant velocity hood is preferred over a bypass hood, which is preferred over a constant volume hood. Whatever hood type is selected, it should have an average face velocity between 0.4 and 0.5 m/s (80–100 ft/min), and other good practices discussed above such as sash position and work practices must be carefully followed.

When using a fume hood (or any other local exhaust system), it is good practice not to directly exhaust effluent that is reasonably suspected to contain ENPs to the outside. The exhaust air should be passed through a high-efficiency particulate air (HEPA) filter (see Section 10.1), since at this time, it is the only air pollution control device known to control nanoparticles with high efficiency.

9.1.3 Alternatives to Conventional Fume Hoods

9.1.3.1 Glove Box   The highest level of protection when handling dry nanopowders is obtained by using a glove-box enclosure (DiBerardinis et al., 1992); a typical glove box is shown in Figure 9.3. Many best practice documents recommend using glove boxes (see, e.g., DOE, 2008; Ellenbecker et al., 2008; NIOSH, 2012; University of Pennsylvania, 2008). The primary advantage of using a glove box is the positive protection it affords; when used properly to manipulate nanopowders, a glove box should reduce the user's exposure close to zero. The disadvantages of using a glove box relate to size limit of the equipment that can be introduced inside the glove box, the extra time required to introduce and remove materials and equipment from the enclosure, the difficulty of manipulating nanomaterials when wearing the bulky gloves, and the need to clean up nanoparticle spills that are likely to occur inside the enclosure. In fact, the two most likely sources of exposure when using a glove box are the transfer of materials into and out of the box, and the cleaning of the box following its use—both of these activities must be performed with extreme care.

9.1.3.2 Biological Safety Cabinet Many best practices documents recommend the biological safety cabinet (BSC) as an improvement to the conventional fume hoods discussed earlier (see, e.g., Ellenbecker et al., 2008; Harvard University, 2011; NIOSH, 2012; Oklahoma University, undated; University of Pennsylvania, 2008). The National Sanitation Foundation (NSF) defines three different classes of BSC:

- A *Class I* BSC resembles a chemical fume hood, with the additional requirement that the exhaust air must be treated before it is discharged to the atmosphere.
- A *Class II* BSC is designed to protect the operator, the product, and the environment. It has an inward air flow through the open sash to protect the operator, a downward flow of HEPA-filtered air to protect the product, and a HEPA-filtered exhaust to protect the environment. Class II cabinets are designed for use against low to moderate risk biological agents.
- A *Class III* BSC is a highly-sophisticated glove box. The sealed enclosure is maintained at a negative static pressure of at least 0.5 in. H_2O, the supply air is HEPA-filtered, and the exhaust air is either double-HEPA filtered or passed through a single HEPA filter and then incinerated. Class III cabinets are highly sophisticated pieces of equipment, meant for the highest-risk biological agents, and are not suitable for working with nanoparticles.

The most widely used class of BSC is Class II; this is the class most likely to be available to researchers when working with nanoparticles. Four different types of Class II cabinets are defined, A, B1, B2, B3; they have differing air flow patterns and increasing levels of control as you move from A to B3.

BSCs are designed to protect personnel working with bioaerosols, such as viruses and bacteria, and are not meant for airborne nanoparticles. However, since airborne nanoparticles behave in a similar fashion like airborne viruses and bacteria, and since the HEPA filtration systems in BSCs should be equally effective in filtering nanoparticles and bioaerosols, it is reasonable to assume that these cabinets will offer similar levels of protection against bioaerosols and airborne nanoparticles. The only available literature (Tsai, 2013) that discusses the use of BSC for nanomaterials concluded that the performance of BSC for containing

(a)   (b)

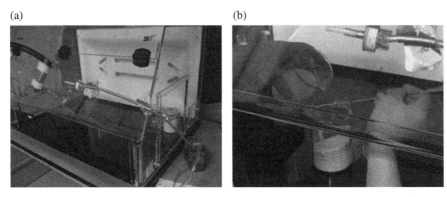

FIGURE 9.9   Powder handling enclosure: (a) general view and (b) close-up of a nanopowder transfer operation. Photos by S. Tsai photos.

airborne nanoparticles was the best when compared to other typical fume hoods such as constant volume, constant velocity, and bypass hoods.

9.1.3.3   Powder Handling Enclosure   Equipment manufacturers now offer ventilated enclosures specifically for weighing and manipulating small quantities of dry powders. These were initially developed and marketed to the pharmaceutical industry but are now sold as general-purpose powder-handling enclosures.[1] Systems typically are self-contained, with their own fan and HEPA filtration unit (an example is shown in Fig. 9.9). The exhaust can be ducted to the outside or recirculated into the room.

A recent publication investigated the effectiveness of powder handling enclosures when handling nanopowders (Tsai, 2013). The results indicated that this type of enclosure is highly effective at containing airborne nanoparticles when handling a typical nanopowder.

Although much more research must be performed to quantify their performance, powder handling enclosures may prove to be the best alternative available when handling small quantities of dry nanopowders. They should work better than conventional fume hoods, be much less expensive than a BSC, and be much easier to work with than glove boxes.

9.1.3.4   Other Ventilation Systems   Some processes that require the manipulation of nanoparticles are too large to fit in a standard laboratory fume hood (e.g., compounding with an extruder). In these cases, an LEV system (Fig. 9.4) can usually be designed to capture the emissions from the process. A detailed discussion of LEV system design is beyond the scope of this book. For more information, it is recommended that one of the standard reference texts be consulted (Burgess et al., 2004; Goodfellow and Tahti, 2001).

---

[1] see for example http://www.a1-safetech.com/ and http://www.labconco.com/_scripts/editc20.asp?CatID=70

It is *extremely important* that the LEV system be designed by a competent professional. The relevant health and safety office should be consulted on the design of a proper system. It is also important that the system be used properly and that it not be modified without consultation with the H&S office.

9.2 CONTROL OF DERMAL EXPOSURES

9.2.1 General

In occupational health, it has been a long-held belief that intact adult human skin is impervious to dry solid particles. Considerable recent research has called this assumption into question for submicrometer-sized industrial aerosols (Tinkle et al., 2003) and ENPs (Rouse et al., 2007). This research indicates that nanoparticles, especially the smallest-diameter particles such as fullerenes (Rouse et al., 2007) and quantum dots (Ryman-Rasmussen et al., 2006), may penetrate the skin. Once such particles enter the skin, there is concern over damage to the skin itself, but also whether they can enter the circulatory system and translocate to other organs. There is evidence that nanoparticle penetration is enhanced if the skin is damaged (Larese Filone et al., 2009).

9.2.2 Clothing and Personal Protective Equipment

Given the possibility that certain ENPs may penetrate the skin, good practice dictates that the skin be protected whenever working with these materials. Basic steps include:

- Wear appropriate personal-protective equipment on a precautionary basis whenever the failure of a single control, including an engineered control, could entail a significant risk of exposure to researchers or support personnel.
- Wear clothing appropriate for a wet-chemistry laboratory including:

  ○ Closed-toed shoes made of a low permeability material. (Disposable over-the-shoe booties may be necessary to prevent tracking nanomaterials from the laboratory.)
  ○ Long pants without cuffs.
  ○ A long sleeved shirt.
  ○ Disposable laboratory coats. If nondisposable laboratory coats are preferred, they should remain in the laboratory/change out area to prevent nanoparticles from being transported into common areas. The coats should be placed in closed bags before being taken out of the laboratory for cleaning in a central approved location.

Wear latex or nitrile gloves when handling nanoparticle powders and nanoparticles in liquids. Atomic force microscopic examination of latex and nitrile gloves in our laboratory found no pores in new, unstretched latex and nitrile gloves, as shown in Figure 9.10, which shows some very thin areas but no actual pores. It should be

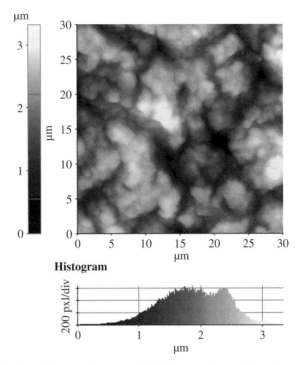

FIGURE 9.10   Atomic force microscope (AFM) image of a new latex glove. Image courtesy E. Ada, University of Massachusetts Lowell.

emphasized that these tests were performed on new gloves; using gloves may cause holes to form, so it is important that gloves be changed frequently. Under severe use conditions, wearing a double layer of gloves is recommended. The contaminated gloves should be kept in a closed plastic bag in the work area until disposal as unregulated hazardous waste.

Outer gloves made of other material, such as cotton, may be used for protection when handling articles wherein the nanomaterials are in bound form.

Safety glasses, and/or face shields, should be worn appropriate based on the level of hazard. A face shield alone is not sufficient protection against unbound dry materials.

9.3   ADMINISTRATIVE CONTROLS AND GOOD WORK PRACTICES

Although engineering controls constitute the first line of defense when working with ENPs, administrative controls are very important as well. Administrative controls are defined as any activities that do not physically change the work environment (i.e., engineering controls) but that contribute to reducing worker exposures. The primary controls as applied to ENPs are described in the following.

9.3.1 Housekeeping

It is important to practice good housekeeping in laboratories where nanomaterials are handled. All working surfaces potentially contaminated with nanoparticles (i.e., benches, glassware, apparatus, exhaust hoods, support equipment) should be cleaned at the end of each day using a HEPA[2] vacuum pickup and/or wet wiping methods. Each laboratory working with nanoparticles in dry form should have access to a HEPA vacuum. The HEPA vacuums should be labeled "For Use with Nanoparticles Only" and used only for this purpose. Dry sweeping or the use of compressed air must be prohibited, since both practices generate aerosols from settled dust.

9.3.2 Work Practices

Good work practices can significantly reduce the potential for nanoparticle release and subsequent exposure. Much of good work practice is common sense, that is, straightforward actions that minimize the risk of nanoparticle release. Many good practices, such as the prohibition of food and drinks in laboratory areas, are common to all laboratories and are well known. The following work practices specific to ENPs are found in many of the reviewed good practices documents.

- Transfer nanomaterial samples between workstations (such as exhaust hoods, glove boxes, furnaces) in closed, labeled containers.
- Do not allow nanoparticles or nanoparticle-containing materials to contact the skin.
- If nanoparticle powders must be handled outside of a ventilated enclosure (see Section 5.2), use appropriate respiratory protection.
- Vacuum up dry nanoparticles only if the vacuum cleaner has a tested and certified HEPA filter.

As an alternative to HEPA-vacuuming laboratory bench tops, bench top protective covering material (e.g., Fisherbrand® Absorbant Surface Liner) can be used; this material should be disposed of as hazardous waste at the end of any day when nano-materials were used on it.

9.3.3 Worker Training

Everyone who works with ENPs should be trained in the risks associated with that work and the proper work practices to be followed to limit exposure. Some best practice documents recommend annual training (Ellenbecker et al., 2008). The US

---

[2] HEPA stands for High Efficiency Particulate Air filter. Shop vacuums equipped with HEPA filters will remove 100% of the nanoparticles in the vacuumed air, whereas many of those particles will pass right through a standard shop vacuum and contaminate the laboratory air.

Department of Energy (DOE, 2008) recommends that training should address, at a minimum, "requirements and recommendations for:

- Employing engineered controls,
- Using personal protective equipment,
- Handling potentially contaminated laboratory garments and protective clothing,
- Cleaning of potentially contaminated surfaces, and
- Disposal of spilled nanoparticles."

Occupational health professionals frequently rely on Material Safety Data Sheets (MSDSs) as an element of worker training. Unfortunately, MSDSs prepared by manufacturers of ENPs have often been found to contain significant shortcomings, particularly in their choice of a reference permissible exposure limit (PEL) (Hodson and Crawford, 2009). CNT exposure can be used as an illustrative example. Since CNTs are essentially pure carbon (with a small amount of metal catalyst), manufacturers in their MSDSs may reference the Chemical Access Service (CAS) number and PEL for carbon black ($3.5\,mg/m^3$) or for graphite (PEL of $2.5\,mg/m^3$) (OSHA, 1992). The *peak* concentration measured by Maynard et al. (2004) at a CNT manufacturing facility was two orders of magnitude lower than these values. Since the health effects associated with CNT exposure likely are related more to the number concentration than the mass concentration, such low exposures on a mass basis may be very significant when converted to number concentration. In 2009, NIOSH reviewed more than 40 MSDSs prepared by ENP manufacturers (Hodson and Crawford, 2009) and reached the following conclusions:

> The quality of information on the MSDSs varied from good to those in need of significant improvement. Some manufacturers had unique disclaimers for their products. Many MSDSs contained the NIOSH-recommended exposure limit, and the OSHA permissible exposure limit for the bulk material (such as graphite) when the material was not graphite, but rather carbon nanotubes. Referencing the bulk chemical exposure limit without any further information may be misleading to the users, since there are published studies demonstrating that the toxicity of ultrafine or nanoparticles is greater than that of the same mass of larger particles of the same chemical composition.

One example will be discussed here. The MSDS for a certain manufacturer of multi-walled CNTs, published on August 29, 2008, and still accessible from the company web site in May 2014, cites the graphite CAS number. Under section 3, Hazards Identification, it fails to list the graphite PEL but instead states "no known hazards, but may cause irritation to eyes, skin, respiratory system, and gastrointestinal tract." Under section 11, Toxicological Information, it states "No known toxicological hazards." Those two statements ignore the CNT toxicology studies reviewed in Section 5.4, many of which were published before August 2008, and many more of which were available when this chapter was written. In spite of those claims of no known toxicity, under section 8, Exposure Controls and Personal Protection, it states "Respiratory Protection: Wear high-efficiency particle respirator to prevent inhalation. Skin Protection: Wear gloves and protective clothing to prevent skin contact."

One technique that we have found to be particularly effective in training our students is the use of fluorescently tagged ENPs. The students perform typical activities, such a weighing and transferring the particles from one beaker to another. Upon completion, as shown in Figure 7.17, their gloves and clothing and the surface where the work was performed show no evidence of contamination under visible light. Under ultraviolet light, however, extensive contamination becomes immediately visible. The lessons are as follows:

- Always wear gloves and protective clothing.
- Always consider gloves and protective clothing to be contaminated and dispose of them accordingly.
- Clean work surfaces on a regular basis, even if they do not appear to be contaminated.

9.4   RESPIRATORY PROTECTION

9.4.1   General Considerations

Respiratory protection should always be viewed as the last line of defense when working with nanoparticles. For both practical and philosophical reasons, the occupational hygiene profession puts respiratory protection at the bottom of the hierarchy of controls. Philosophically (and legally, in many jurisdictions), occupational health professionals believe that it is the responsibility of the *employer* to provide the *employee* with a safe place to work; wearing a respirator turns this around and puts the burden on the worker. The practical reasons are that respirators are uncomfortable to wear, make respiration and communication more difficult, and may interfere with the work that needs to be performed. In addition, respirators are expensive, since they must be used as part of a completer respiratory protection program, as described below.

In the United States, this approach to respiratory protection is codified by paragraph (a)(1) of OSHA's respiratory protection standard (OSHA, 1998):

> In the control of those occupational diseases caused by breathing air contaminated with harmful dusts, fogs, fumes, mists, gases, smokes, sprays, or vapors, the primary objective shall be to prevent atmospheric contamination. This shall be accomplished as far as feasible by accepted engineering control measures (for example, enclosure or confinement of the operation, general and local ventilation, and substitution of less toxic materials). When effective engineering controls are not feasible, or while they are being instituted, appropriate respirators shall be used pursuant to this section.

It is important to distinguish between mandatory and voluntary respirator use. Under the Occupational Safety and Health Act, respirator use is *required* only if worker exposure to some contaminant exceeds the PEL (or, in limited cases, an exposure level lower than the PEL called the action level). Almost all aerosol PELs are based on mass concentration. Since there are no PELs specifically for ENPs, and since exposures to nanoparticles typically are very low on a mass basis, it is unlikely that any workers in a nanoparticle research laboratory would be exposed to airborne nanoparticles at a level that *requires* respiratory protection. On the other hand, workers handling ENPs may *choose* to wear a respirator voluntarily.

9.4.2 Respirator Designs

9.4.2.1 Respirators for Nanoparticle Collection

Any respirator designed for removing aerosols from the breathing air, no matter what basic design, relies on filtration to collect the particles. As discussed in Chapters 7 and 9, filters have been used as air sampling and air cleaning devices for more than 50 years; over that time, very effective respirator filters have also been developed. A filtering respirator must meet two primary design goals: remove aerosol particles from the breathing air with high efficiency and have a relatively low breathing resistance, or static pressure drop. As we shall see, these factors typically are in conflict in that techniques to increase filter efficiency also increase pressure drop; this presents significant design challenges to respirator manufacturers.

There is a well-developed theory to describe the collection efficiency and pressure drop of filters; these theories will be presented for particle collection and pressure drop across fibers, since most respirator filters use fibers as their collection medium.

Collection Efficiency   Collection efficiency is defined as the fraction of particles entering a respirator filter that is removed by that filter. We can define both mass collection efficiency, η_m, and particle number collection efficiency, η_n:

$$\eta_m = \frac{c_{in} - c_{out}}{c_{in}} \tag{9.2}$$

$$\eta_n = \frac{N_{in} - N_{out}}{N_{in}} \tag{9.3}$$

where c_{in} = aerosol mass concentration entering the collection device (mg/m^3), c_{out} = aerosol mass concentration leaving the collection device (mg/m^3), N_{in} = aerosol number concentration entering the collection device (particles/cm^3), and N_{out} = aerosol number concentration leaving the collection device (particles/cm^3).

Figure 9.11 shows the cross-section of a fiber and the air flow streamlines passing around that fiber. In the bulk air flow, particles are carried along on the streamlines at the same velocity as the air. In order for particles to collect on a fiber, they must leave their streamline (with one exception) and contact the fiber; the figure illustrates the five principal methods by which such collection occurs.

- In *inertial impaction*, the particle's forward momentum (see Chapter 2 for a discussion of particle inertia) causes it to tend to continue to move straight ahead as the streamline occurs. If this momentum is sufficiently strong, the

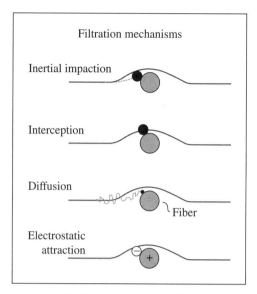

FIGURE 9.11 Cross-section of a fiber, air flow streamlines, and particle collection mechanisms. Diagram courtesy of U.S. NIOSH (NIOSH 2003).

particle will strike the fiber. Impaction is enhanced for large particles and high filtration velocities. It is also enhanced by using smaller diameter fibers, since streamlines approach closer to the fiber surface before turning, thus requiring particles to travel shorter distances in the forward direction.

- With *interception*, particles do not have to leave their streamline; if a particle is traveling on a streamline that passes within its radius of the fiber surface, it will strike the surface and be collected. Interception is favored by larger particles.

- *Diffusion* refers to the collection of particles due to their Brownian motion (also defined in Chapter 2). Small particles randomly move off their streamlines due to Brownian motion, and as the streamline passes a, fiber there is some probability that the particle will strike the fiber and be collected. This is termed diffusion since particles make a net motion toward the fiber without other particles moving back. Diffusion is favored by smaller particles, which have higher Brownian motion, and lower filtration velocities, since this increases the particle's residence time in the vicinity of the fiber.

- *Gravity settling* (not shown in Fig. 9.11), also termed sedimentation, was discussed in Chapter 2. It usually plays a very small role in filter particle collection.

- Finally, *electrostatic charge* can be used to enhance the collection efficiency of charged particles by introducing excess charges on the filter material during the manufacturing process. Freshly produced aerosol particles can have a significant positive or negative charge. As the aerosol ages, it encounters positive and negative air ions continuously produced by cosmic radiation, which tend to reduce the

TABLE 9.1   Distribution of charge on aerosol particles at boltzmann equilibrium

Particle diameter (nm)	Average number of charges	Percentage of particles carrying the indicated number of charges						
		<-2	-2	-1	0	+1	+2	>+2
10	0.007			0.3	99.3	0.3		
20	0.104			5.2	89.6	5.2		
50	0.411		0.6	19.3	60.2	19.3	0.6	
100	0.672	0.3	4.4	24.1	42.5	24.1	4.4	0.3
200	1.00	2.6	9.6	22.6	30.1	22.6	9.6	2.6
500	1.64	11.4	12.1	17.0	19.0	17.0	12.1	11.4
1000	2.34	19.9	10.7	12.7	13.5	12.7	10.7	19.9
2000	3.33	27.5	8.5	9.3	9.5	9.3	8.5	27.5

Adapted from Hinds (1999).

excess charge to a fairly low level. Even well-aged aerosols have a probability of having a few excess charges; the probability of different excess charge numbers differs by particle size, as given by the Boltzmann distribution. Table 9.1 shows the Boltzmann charge distribution for different particle diameters (Hinds, 1999). As charged aerosol particles pass by charged fibers, oppositely charged particles will be attracted to the fiber, enhancing particle collection.

When a gas (for respirators, air) passes through a filter, individual particles may be removed by one of the above collection mechanisms, where the effectiveness of each mechanism depends on the factors discussed above. Equations can be developed for the single-fiber collection efficiency of each mechanism, defined as the fraction of particles originally on streamlines approaching the fiber cross-sectional area that are removed by that mechanism (Fig. 9.11). Although the factors contributing to the collection efficiency of each mechanism are well understood, the equations themselves are empirical (Hinds, 1999). Once the single-mechanism single-fiber collection efficiency is determined, they can be combined to determine the total single-fiber collection efficiency, η_s, which typically is a strong function of particle size. Finally, the total filter collection efficiency is given by

$$\eta = 1 - e^{-4\alpha\eta_s t / \pi d_f} \tag{9.4}$$

where α = filter solidity (fraction of the filter volume taken up by fibers), t = filter thickness (m), and d_f = fiber diameter (m).

The total collection efficiency of a typical filter as a function of particle size at a given velocity is shown in Figure 9.12. Starting from the left of the curve, very small particles have a high Brownian motion, and thus are removed efficiently by diffusion. As the particles become larger, diffusion is less effective and the curve drops. Large particles are collected efficiently by inertial impaction and interception, so the curve starts climbing again and approaches 1.0 for the largest particles. As shown in the

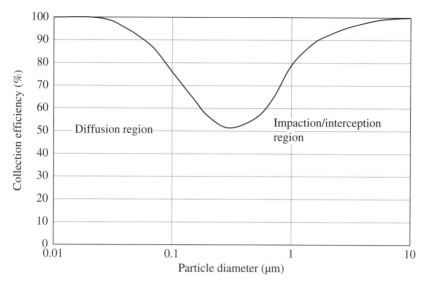

FIGURE 9.12   Total collection efficiency of a typical filter as a function of particle size.

TABLE 9.2   Performance criteria for NIOSH-certified respirator particulate filters

Oil resistance	Rating	Performance
Not oil resistant	N95	Filters at least 95% of airborne particles
	N99	Filters at least 99% of airborne particles
	N100	Filters at least 99.97% of airborne particles
Oil resistant	R95	Filters at least 95% of airborne particles
	R99	Filters at least 99% of airborne particles
	R100	Filters at least 99.97% of airborne particles
Oil proof	P95	Filters at least 95% of airborne particles
	P99	Filters at least 99% of airborne particles
	P100	Filters at least 99.97% of airborne particles

figure, there is an intermediate range of particle diameters for which none of these three mechanisms are effective; this size typically lies at about 200–400 nm. It is for this reason that filter efficiency test methods typically specify the use of test aerosols in this size range.

Applying this to the United States, the National Institute for Occupational Safety and Health (NIOSH) certifies particulate respirator filters of the types shown in Table 9.2, based on use and minimum collection efficiency. The NIOSH Respirator Rule, 42 CFR Part 84 (NIOSH, 1995), states that "all filter tests will employ the most penetrating aerosol size, 0.3 μm aerodynamic mass median diameter." The European Union has issued similar standards for respirators used in their member countries, as shown in Tables 9.3 (European Commission for Standardization 2006) and 9.4 (European Commission for Standardization 2010).

TABLE 9.3   Performance criteria under european standard
EN 143 for respirator filters that can be applied to a face mask
(european commission for standardization, 2006)

Class	Filter penetration limit (at 95 l/min air flow)
P1	Filters at least 80% of airborne particles
P2	Filters at least 94% of airborne particles
P3	Filters at least 99.95% of airborne particles

TABLE 9.4   Performance Criteria Under European Standard EN 149
for Filtering Facepiece Respirators (European Commission for
Standardization, 2010)

Class	Filter penetration limit (at 95 l/min air flow)	Inward leakage
FFP1	Filters at least 80% of airborne particles	<22%
FFP2	Filters at least 94% of airborne particles	<8%
FFP3	Filters at least 99% of airborne particles	<2%

Pressure Drop   The second most important factor for characterizing a respirator filter is the static pressure loss across the filter, commonly called the pressure drop and typically measured in millimeters water column (mm H_2O). In a fibrous filter, increasing the filtration velocity, the filter thickness and/or the solidity will cause an increase in pressure drop; decreasing the fiber diameter will also increase pressure drop.

While efficiency is a measure of the *effectiveness* of a respirator filter, pressure drop is a measure of the *energy* required from the wearer to pull air through the cartridge. An ideal respirator filter would have 100% collection efficiency and zero pressure drop, and an actual filter's performance is measured against these ideals. In a respirator, the pressure drop across the filter elements is especially important, since the energy to overcome it is provided by the respirator wearer when (s)he breathes. OSHA requires that NIOSH-certified air-purifying particle-removing respirators have an inhalation pressure drop no greater than 35 mm (1.4 in.) H_2O at a continuous air flow of 85 L/min (NIOSH, 1995).

Filter Design   The problem in designing an air-purifying particle-removing respirator is to meet *both* the required minimum collection efficiency and the required maximum pressure drop. Let's assume that the original effort to design an N95 respirator resulted in a respirator that just met the pressure drop requirement with the filter collection efficiency curve of Figure 9.13; clearly, the collection efficiency falls below 95% for particles of about 300 nm. The designer has several options available to increase the filter collection efficiency. First, he could increase the thickness of the filter; as indicated in Equation (9.4), this will increase the overall collection of the filter—in effect, raising the dip in the curve to the required 95% level. However, this has the added negative effect of increasing the respirator's breathing resistance.

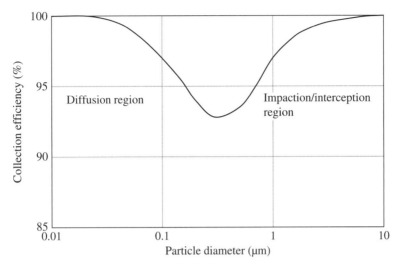

FIGURE 9.13   Filter collection efficiency curve, N95 filter with no electrostatic charge. Note that the vertical axis starts at 85%.

In considering Equation (9.4), the other possible design changes that could increase efficiency are increasing the filter solidity and/or decreasing the filter fiber diameter, but both of these steps also increase breathing resistance.

Given these inappropriate choices, the N95 respirator designer is left to rely on the electrostatic effect to attain the desired collection efficiency at an acceptable pressure drop. In this approach, the filter medium is imparted an electric charge during the manufacturing process—the most common filter of this design is the electret filter (Hinds, 1999). As discussed earlier, aerosol particles typically have a small excess charge; respirator manufacturers have found that the use of a charged filter material can give a "boost" in the collection efficiency of up to 40%, allowing the filter shown for example in Figure 9.14 to now pass the certification test without having to increase the breathing resistance (Jasper et al., 2007).

Recently, N95 respirators have been evaluated for their ability to collect particles in the nanometer-size range (Rengasamy and Eimer, 2012; Rengasamy et al., 2011). Interestingly, although the tested respirators generally had collection efficiencies just about equal to the required 95% at the most penetrating particle size, that size was not at the expected 200–300 nm but rather at about 30 nm. The reason can be seen from an examination of Table 9.1, which indicates that particles in this size range are not likely to have an excess charge and thus would not be collected by electrostatic attraction. Particles smaller than this size also would not likely have a charge but have higher Brownian motions and thus are efficiently collected by diffusion. The key point to stress is that N95 respirators appear to allow about 5% particle penetration, as expected, but that the penetrating particles are well within the nanoparticle size range. This means that wearers of this type of respirator may still have a significant ENP exposure.

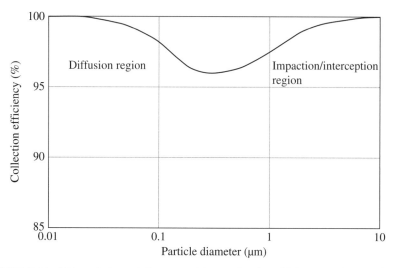

FIGURE 9.14   Effect of charged filter material on collection efficiency on the N95 filter of Figure 9.13.

The N-, R-, and P100 (Table 9.2), PP (Table 9.3), and FFPP (Table 9.4) respirators typically utilize cartridges containing HEPA filters. A typical HEPA filter is constructed of a filter paper that is folded within the filter housing to give it a very high surface area. From Equation (9.1), we know that if a filter's area is increased, the filtration velocity must decrease; thus, one characteristic of a HEPA filter is a very low filtration velocity. This has two primary advantages, the first being that the pressure drop across the filter is very low. The second has to do with overcoming the difficulty in a typical filter of collecting particles in the intermediate size range, typically around 0.2–0.3 μm. As discussed earlier, small particles are collected primarily by Brownian motion and large particles by inertial impaction and interception; none of these mechanisms are very effective in the 0.2–0.3 μm range. Lowering the filtration velocity has the effect of increasing the particle residence time in the filter and allows Brownian diffusion to be effective in this size range. Lowering velocity decreases collection efficiency by impaction, but has minor effects on interception, so that the upper half of the curve does not shift; the result is the almost-flat HEPA collection efficiency curve shown in Figure 9.15. Recent research has confirmed that respirators fitted with HEPA-type filters are highly effective in filtering nanoparticles (Rengasamy et al., 2008)

Respirator Fit   The respirator collection efficiency as discussed earlier only applies to the actual respirator filter; unfortunately, this does not represent the only pathway for ENP penetration of a respirator. It has been known for many years that the *fit* of a respirator strongly influences its performance, since the process of breathing in while wearing a negative-pressure respirator can draw air in between the respirator facepiece and the skin (Popendorf, 2006). The overall performance of a respirator, taking into account both filter performance and facepiece leakage, is given by its

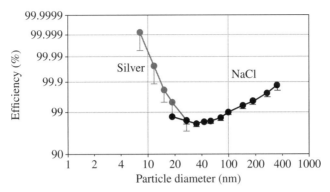

FIGURE 9.15   Collection efficiency of HEPA filter. Note that the vertical axis starts at 90%. Diagram courtesy of U.S. NIOSH.

FIGURE 9.16   Half-mask respirator. Photo courtesy of U.S. NIOSH.

protection factor, defined as the ratio of the contaminant concentration *outside* the respirator to the concentration *inside* the respirator. There are several definitions of different protection factors, including the assigned protection factor (APF) which is determined by an agency such as NIOSH, which certifies respirators in the United States, and the workplace protection factor, which is the actual protection factor attained by a worker wearing the respirator while working. There are no studies in the literature reporting on the actual level of face seal leakage of ENPs, so the conservative approach at this time is to rely on consensus APFs when choosing a respirator.

The most commonly worn respirator, besides the disposable dust mask, is the half-mask respirator (Fig. 9.16); for protection against ENPs, it can be fitted with an N or P95 or 100 filter cartridge. The NIOSH APF for this respirator is only 10, meaning

FIGURE 9.17   Full facepiece respirator. Photo courtesy of U.S. NIOSH.

that it typically only reduces the concentration by a factor of 10, which corresponds to a penetration of 10%. Since the filter is either at least 95% efficient (N, P95) or close to 100% efficient (N, P100), it is clear that most of the penetration into this type of respirator is by face seal leakage. Full facepiece respirators (Fig. 9.17) offer better fit to most wearers and thus have a higher protection factor. The NIOSH APF for a full-face respirator with a N100 or P100 cartridge is 50; due to the higher filter penetration, the APF when fitted with an N95 or P95 cartridge is only 10.

If higher protection factors are required, workers must wear a respirator that provides positive pressure inside the facepiece; in this case, any face seal leakage will be out rather than in. Positive pressure respirators, including self-contained breathing apparatus, supplied airline respirators, and powered air purifying respirators, can have APFs ranging from 50 to 10,000, depending on specific design features (NIOSH, 1987).

Recommendations   The following guidelines should be followed:

- The appropriate respirator and cartridge combination, based on an EHS analysis, should be worn when deemed necessary by a safety assessment. The type of respirator selected for use should be sufficient to reduce ENP exposures to the background level, or lower.
- Personnel required to wear a respirator must do so as part of a complete respiratory protection program, as outlined by OSHA (1998), including medical clearance by medical doctor before being fitted with a respirator, quantitative fit testing, and so on. If a respirator is required, it should be at a minimum a half-mask, N-100 or P-100 cartridge-type respirator that has been properly fitted to the worker.
- Personnel not required to wear a respirator may do so at their discretion. In this case, disposable respirators (also called dust masks—surgical masks should be avoided) with at least an N-95 filter rating are acceptable.

Many research laboratories have their own respirator program that is more specific than these general recommendations—contact the applicable EHS office whenever respirator use is contemplated.

9.5   CASE STUDY: COMPARISON OF THE PERFORMANCE OF VARIOUS FUME HOODS

As described in this chapter, several different types of laboratory fume hoods are available and used in both occupational and laboratory settings. Those fume hoods are designed with various features that are incorporated as the technology is evolving. The weighing tests discussed in Chapter 8 showed significant particle release during the weighing process and demonstrated that exposure could occur while handling nanoparticles due to nanoparticle escape from a constant-volume hood. Our studies have demonstrated that airflow pattern plays the most important role to contain nanoparticles, so how would the airflow pattern be affected and vary for different hood designs? How effective would the performance be when handling nanoparticles in different types of hoods? This case study provides a comparison of the effectiveness when using various typical hoods including constant-volume, bypass, and constant-velocity hoods.

> *Case theme:* Manual transferring or pouring nanoparticle powder in three typical types of fume hoods.
>
> *Equipment:* 100 g of nanoparticles, beakers, spatulas, fog generator, a green-light laser, and three types of fume hoods.
>
> *Investigation:* Effectiveness of nanoparticle containment in typical fume hoods.

9.5.1   Materials and Hoods

Aluminum oxide[3] (nanoalumina) nanoparticles were handled to demonstrate weighing tasks, where 100 or 15 g of nanoparticles were transferred using a spatula or poured from beaker to beaker. The test nanoalumina particles appear roughly spherical in shape with a primary particle size ranging from 27 to 56 nm; when dried, these particles formed agglomerates in the bulk material with a nominal size of about 200 nm.

Particle handling was studied using a constant-volume (conventional) hood, a bypass hood, and a constant-velocity hood (ACGIH, 2007). With no special features to help control the hood face velocity, a constant-volume hood has a constant air flow, and the hood face velocity varies inversely with the height of the sash opening. The studied conventional hood (Fig. 9.18) has full open sash face dimensions of 62 cm (24 in.) high × 130 cm (51 in.) wide.

---

[3]Also called nanoalumina, grade Al-015-003-025, they were manufactured using physical vapor synthesis (PVS) by Nanophase Technologies Corporation, Romeoville, IL, http://www.nanophase.com/technology/

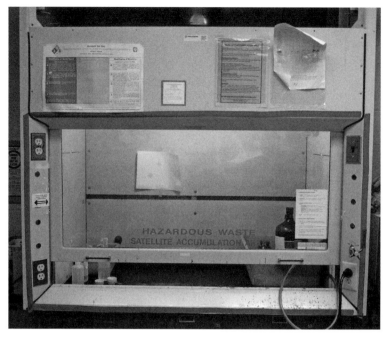

FIGURE 9.18 Constant-flow fume hood. Manufacturer: Fisher Scientific Safety-Flow Laboratory Fume Hood, Model 93-509Q. Photo by S. Tsai.

A bypass hood attempts to maintain a constant velocity through the use of a bypass grille located above the hood opening and attached to the hood exhaust fan. When the sash is wide open, it blocks the bypass grille, and as the sash is lowered the grille is uncovered, allowing increasing amounts of air to flow through the bypass grille instead of the hood face, thus helping to maintain a constant face velocity. The full open sash face dimensions of the studied bypass hood (Fig. 9.19) are 71 cm (28 in.) H × 218 cm (86 in.) W.

A constant velocity hood, also called a VAV hood, uses a motor controller to vary the fan speed as the sash is moved in order to maintain a constant hood face velocity. The sash of the studied constant velocity hood (Fig. 9.20) has dimensions of 69 cm (27 in.) H × 163 cm (64 in.) W.

These three hoods were installed and used in three different operational laboratories, so that the layout and setting in each laboratory are different from each other. Table 9.5 lists some of the important operating characteristics of the hoods at the time of their evaluation; as indicated, measurements were taken at conditions of similar relative humidity.

Airflow surrounding the sash opening can affect the air stream flowing across sash opening into the hood; such external airflows are commonly called crossdrafts. Crossdraft velocities were detected in the vicinity of both the bypass and constant velocity hoods (Table 9.5). The crossdraft at the bypass hood was caused by an air conditioner installed on the side wall which caused the air flow across the open sash

FIGURE 9.19 Bypass hood. Photo by S. Tsai.

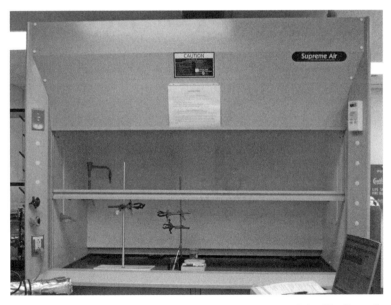

FIGURE 9.20 Constant-velocity hood. Manufacturer: Kewaunee Scientific Corp., Supreme Air Model H05, with a Phoenix Controls Corp. flow control system. Photo by S. Tsai.

with a velocity as high as 0.18 m/s (35 ft/min). The crossdraft at the constant velocity hood, up to 0.13 m/s (25 ft/min), was less intense than the cross flow at the bypass hood. Conditioned air supplied from the ceiling, located in front of the constant velocity hood, caused air to flow periodically across the open sash.

TABLE 9.5   Hood operating profiles

Hood type	Years used	Cross draft m/s (ft/min)	Minimum face velocity m/s (ft/min)	RH (%)
Conventional	>20	0	0.4 (80)	48–52
By-pass	<5	Up to 0.2 (35)	0.3 (70)	50–60
Constant velocity	<1	Up to 0.1 (25)	0.5 (100)	63

9.5.2   Measurements

Particle handling was performed following the same procedure described in Section 8.4. Particles were transferred or poured in each hood. Measurements were taken at the researcher's breathing zone and at the source upstream and downstream side locations. Particle release measurements and equipment used for this study are described in Section 8.4. In addition to using the Fast Mobility Particle Sizer (FMPS) to measure nanoparticles, the concentrations of airborne particles in the micrometer size range (diameters from 0.5 to 20 μm) were measured using the Aerodynamic Particle Sizer (APS®) spectrometer (Model 3321, TSI). Nanoparticle-contaminated surfaces inside the hood were cleaned after completion of all handling tasks, and particle release during the cleaning task was monitored. The contaminated surfaces were cleaned using wet paper towel saturated with water to wipe up the spilled nanoparticle powder.

A fog generator and a green-light laser producing a laser-light sheet with 500 mW at peak power and wavelength of 473–671 nm (Laserglow) were used for visual tests of airflow patterns. A 163 cm height female mannequin with chest width of 25 cm was placed in front of each hood, and the two arms of the mannequin were positioned inside the hood at the workers' manual handling position (Fig. 9.21).

9.5.3   Results

The particle releases to worker's breathing zone during particle handling in these three hoods are compared. The breathing zone exposure concentration of the constant-flow hood is shown in Figure 8.5c and d. The particle concentrations at breathing zone of bypass and constant-velocity hoods during handling particles are shown in Figures 9.22 and 9.23, respectively.

The results while using the constant-flow hood, as seen in Figure 8.5c d, showed the highest breathing zone concentration for all three sash locations and for both the transferring and pouring tasks compared to the concentration measured using constant-velocity and bypass hoods. The bypass hood provided performance in between the constant-flow and constant-velocity hoods, and high-sash operations caused higher exposure for both transferring and pouring tasks as shown in Figure 9.22. For the constant-velocity hood, particle release was barely detected while transferring nanoparticles (Fig. 9.23). However, the pouring method resulted in a more intense nanoparticle exposure as seen in Figure 9.23b, the concentration peak was found at approximately 200 nm particle diameter.

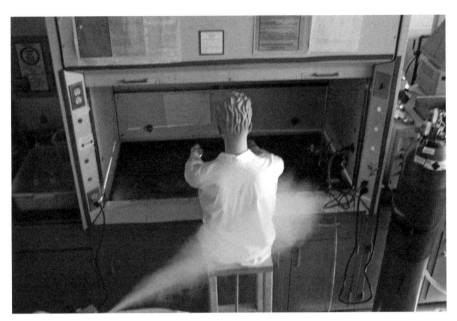

FIGURE 9.21 Experimental setup of airflow examination using fog, laser, mannequin, and a constant-volume hood. Note: Fog can be generated inside and outside the fume hood. Photo by S. Tsai.

9.5.3.1 Effect of Sash Location The sash location affects the face velocity of the constant-flow hood and also the constant-velocity hood when it is fully open. Nanoparticle release was affected by changes in face velocity, as shown in Figure 9.23. For the constant-flow hood, the highest exposure occurred at the low sash position for both the transfer and pouring tasks. The released particles were predominantly larger than 100 nm diameter (Figure 8.5c and d). The breathing zone concentration decreased when the face velocity was reduced; the lowest exposure was measured at the high sash location which had a face velocity of 0.4 m/s (80 ft/min). The middle sash location, which is the normal operating position, provides a face velocity of 0.6 m/s (120 ft/min) and results in a higher concentration for particles less than 50 nm diameter.

For the constant-velocity hood, an increase in nanoparticle breathing zone concentration was barely detected when operating at the desired face velocity of 0.5 m/s (100 ft/min) while at the normal operating position (middle sash) or low sash for both the transfer and pouring tasks (Figure 9.23a and b). The exposure became unstable at the high sash, where the face velocity dropped to 0.3 m/s (60 ft/min). More particles were released during the pouring task, which is more energetic than transfer.

Since the nanometer-sized nanoalumina particles may agglomerate in the bulk powder material, limited experiments were also performed using the APS. Particle release for diameters from 5 nm to 20 μm, using the single operation of pouring for each hood and the same operator, is shown in Figure 9.24. Using the more energetic task of pouring, the release of particles larger than 500 nm diameter was barely detectable; in other words, particles escaping from the hoods during the handling of

(a)

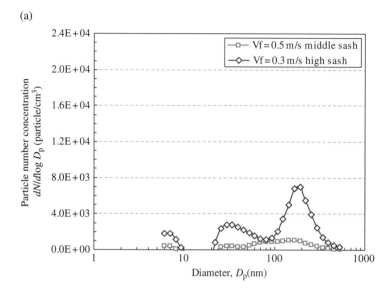

(b)

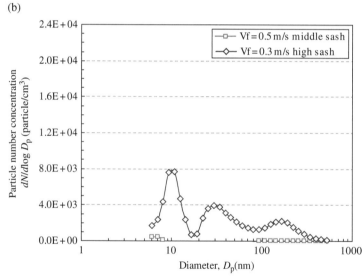

FIGURE 9.22   Increase in the breathing zone concentration during handling 100 g nanoalumina particles in the *bypass hood*: (a) transferring and (b) pouring. Reprinted with permission from Tsai et al. (2009b). Copyright 2009 *Journal of Nanoparticle Research*.

nanoalumina particles are almost all smaller than 500 nm. The results of the tests in Figure 9.24 are consistent with the averaged performance over various conditions shown in Figures 8.5, 9.22, and 9.23. The highest particle concentration escaping from the hoods still occurred in the constant-flow hood. The release was better controlled by using the constant-velocity hood. The set of operations shown in Figure 9.24 were performed by an operator of small stature, with a chest width of about 34 cm

(a)

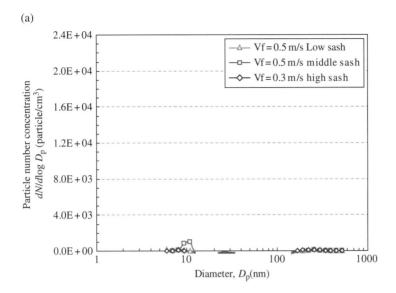

(b)

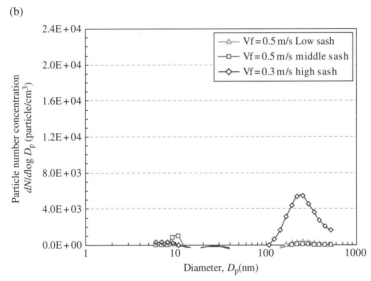

FIGURE 9.23   Increase in the breathing zone concentration during handling 100 g nanoalumina particles in the *constant-velocity hood*: (a) transferring and (b) pouring. Reprinted with permission from Tsai et al. (2009b). Copyright 2009 *Journal of Nanoparticle Research*.

(13 in.) and a height of 174 cm (68 in.), and very slow and gentle motions were applied through all handling tasks. The high concentration peaking at about 200 nm diameter shown for the constant-flow and constant-velocity hoods in Figures 8.5 and 9.23 is not seen in Figure 9.24. For a lower release from this set of operations, the optimal handling condition with lowest particle release occurred at the face velocity

(a)

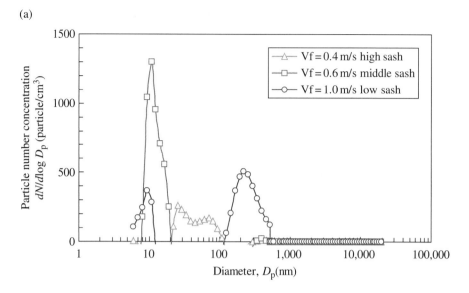

(b)

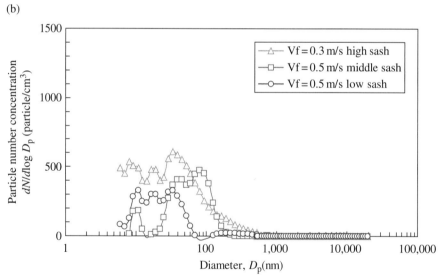

FIGURE 9.24 Particle number concentration increase (5–20,000 nm) during pouring nanoparticles during one pouring operation. (a) Constant-flow hood and (b) constant-velocity hood. Used by permission of (Tsai et al. 2010). Copyright 2010 Oxford University Press.

of 0.4 m/s (80 ft/min) (high sash) for the constant-flow hood (Fig. 9.24a) and at the face velocity of 0.5 m/s (100 ft/min) (middle, low sash) for the constant-velocity hood (Fig. 9.24b). Handling nanoparticles in the constant-flow hood at face velocities between 0.4 and 0.5 m/s (80–100 ft/min) resulted in the lowest exposure under most operating conditions studied.

9.5.3.2 Air Flow Patterns   The results of the flow visualization tests of airflow patterns for the three hoods are shown in Figure 9.25. Photographs were taken at the middle sash locations for all hoods. For the constant-flow and constant-velocity hoods, which generate an air flow past the operator, fog was generated behind the mannequin; the airflow carried the fog flow around the mannequin and into the hoods. For the constant-flow hood, horizontal and vertical views are shown in Figures 9.25a and b, respectively. Circulating airflows in the horizontal plane were intense and airflow eddies are readily observed between the mannequin's body and arms (Fig. 9.25a). The vertical view of airflow at a plane 15 cm behind the doorsill with arms inside the hood duplicating the actual position of handling tasks described earlier shows very intense airflows circulating individually around each arm of the mannequin (Fig. 9.25b). These vertical circulations around the arms were seen only in the constant-flow hood. For the constant-velocity hood, the airflow seen in the vertical plane (Fig. 9.25d) did not show the individual circulating airflow around the two arms, instead, a downward airflow was seen to evenly carry smoke around the arms. Circulating airflows were observed in the horizontal plane in the constant-velocity hood, but the eddies formed between the mannequin's body and arms were smaller than for the constant-flow hood and the circulation was less intense (Fig. 9.25c).

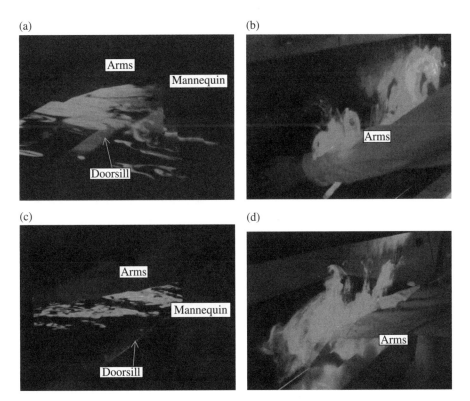

FIGURE 9.25   Features of air flow pattern of constant-flow and constant-velocity hoods at middle sash. (a) Constant-flow hood, horizontal; (b) constant-flow hood, vertical; (c) constant-velocity hood, horizontal; and (d) constant-velocity hood, vertical. Photos by S. Tsai.

9.5.3.3 Particle Accumulation from Hood Leakage The change in total concentration at the worker's breathing zone during handling nanoparticles is plotted as a function of time in Figure 9.26, after subtracting the breathing zone concentration at time $t=0$ just before handling particles began. The four measurements were taken at a relative humidity of 48–63% (Table 9.5). The concentrations increased with time for experiments using 100 g handling in the constant-flow and bypass hoods, while the concentrations during handling 15 g nanoalumina particles in the constant-flow and constant-velocity hoods remained below the baseline concentration during handling tasks as shown by the negative data in Figure 9.26. Nanoparticle exposure was significant for all measurements during handling particles in the 20-year-old constant-flow hood. For handling 100 g nanoalumina in the bypass hood, the concentration increase was less than that measured in the constant-flow hood. The 15 g handling tasks performed in the constant velocity hood did not generate accumulated particles as detected by the FMPS, but a concentration increase while handling in the constant-flow hood can be seen between 10 and 20 min after handling 15 g nanoalumina.

Results indicate that the breathing zone particle concentration was the highest when working in the constant flow hood while handling 100 g nanoalumina. For the 100 g handling task, particle concentration at the worker's breathing zone accumulated due to excess nanoparticles that were generated inside the hood and carried out of the hood resulting in a dramatic increase of nanoparticles at worker's breathing

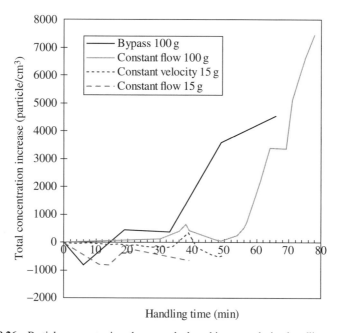

FIGURE 9.26 Particle concentration change at the breathing zone during handling nanoalumina particles in the constant-flow, bypass, and constant-velocity hood. Reprinted with permission from Tsai et al. (2009b). Copyright 2009 *Journal of Nanoparticle Research.*

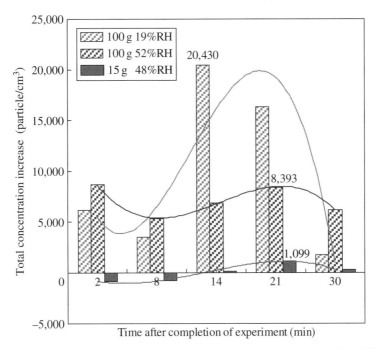

FIGURE 9.27   Particle concentration change at the background after handling 100 g and 15 g nanoalumina particles in the constant flow hood. Reprinted with permission from Tsai et al. (2009b). Copyright 2009 *Journal of Nanoparticle Research.*

zone. This also results in an increased particle concentration in the laboratory air right after completing handling experiments as can be seen in Figure 9.27.

Cleaning the contaminated work surface inside the hood was the main task performed after completion of the nanoparticle handling experiments. Concentrations in the general laboratory air after completing handling tasks increased greatly while cleaning spilled particles at the hood workplace. Three experiments at different environmental conditions were performed in the constant flow hood and the results showed that particle concentrations were reduced slightly at 8 min after completing experiment and rose quickly during cleaning between 14 and 21 min after experiment; concentrations declined after cleaning was complete. Particle concentration during cleaning in a low relative humidity (19%) environment was much higher than in a high relative humidity (52%) environment for 100 g handling experiments. The 15 g handling experiment yielded significantly lower particle release into the laboratory air.

The magnitude of particle release and exposure to the worker could vary over a broad range and was affected by the hood design, handling tasks, types of materials, quantity, and so on. All results discussed thus far were for nanoalumina; experiments with nanosilver found that more nanoalumina particles were entrained into the displaced air stream compared to the number of nanosilver particles under the same experimental conditions. Because of this, compared to nanosilver, nanoalumina particles were carried out in dramatically larger numbers into the air inside the hood and recirculated in the vortex. The frequent dynamic motions of handling nanoparticles in the wake region carried nanoparticles out to the worker's breathing zone, and this effect was much more pronounced for nanoalumina than for nanosilver.

For typical hoods being used in laboratory or occupational settings, these experimental results indicate that a likely exposure would occur under certain circumstances. What other control methods or alternatives could be applied for reducing exposure?

9.5.4 The Challenge and Brainstorming

When Sophia walked in the laboratory after lunch to continue her work, she saw a new instrument placed inside the constant-velocity hood which she usually used for preparing her samples, weighing and mixing various metal oxide nanoparticles with solvent. This instrument occupied about half of the hood space; Sophia heard from the principal investigator that this instrument will stay in the hood for a couple of weeks for a pilot project to synthesize CNTs. She has no other hood available for her to perform her work; so she tried to use the remaining space in the hood. Sophia kept thinking about the instrument sitting in the hood; she was worried and wondered "will this cause a problem such as exposure to me either from the CNT synthesis or the nanoparticle handling tasks in the half hood space?"

Sophia would like to discuss this with her boss; however, she did not know what reasoning or scientific evidence that she could use to bring her concerns and to propose an engineering solution to this situation.

9.5.5 Study Questions

1. What is the main reason to cause exposure to workers from using these three types of hoods?
2. What factors dominate the difference in hood performance?
3. When regular fume hoods are unavailable, what engineering controls can be considered?
4. In comparing the potential exposure from powder handling task and particle release from CNT synthesis that Sophia questioned, what additional controls can be applied with the lowest cost to avoid exposure in this case?

9.6 CASE STUDY: PERFORMANCE OF NONTRADITIONAL FUME HOODS

Unique applications or particular needs frequently are the driving force to promote the customized design of local exhaust systems, including laboratory hoods. For those typical laboratory fume hoods described in Section 9.1.2, when they could not effectively contain airborne nanomaterials under certain operating conditions, would other exhaust systems or hoods work better? What hoods or new designs are available for nanotechnology relevant activities? This case study presents evaluations of several different laboratory hoods for the performance of containing airborne nanoparticles. The studied hoods include the biological safety cabinet (BSC), powder handling enclosures, and the air-curtain hood.

TABLE 9.6   Hood Specifications and Operating Conditions, Second Case Study

	Class/Width, cm (inch)	Sash	Face velocity, m/s (ft/min)	Front exhaust velocity, m/s (ft/min)
BSC 1	Class II B2/BSL 178 (70)	Movable sash (half/full open)	0.30–0.38 (60–75)	2.3 (460)
BSC 2	Class II A/B3 119 (47)	Fixed sash (center)	0.13–0.46 (25–90)	2.7 (540)
Powder handling enclosure 1	91 (36)	Fixed sash	0.38 (75) (0.38–0.51 tested)	N/A
Powder handling enclosure 2	91 (36)	Fixed sash	0.37 (73) (0.26–0.37 tested)	N/A
Air-curtain hood	152 (60)	Movable sash	N/A	Air curtain flow

Case theme: Manual transferring or pouring nanoparticle powder and visually examining air flow pattern in five different hoods.

Operation: 100 g of nanoparticles, beakers, spatulas, fog generator, a green-light laser, and five hoods.

Investigation: Effectiveness of nanoparticle containment in nontypical hoods.

9.6.1   Materials and Hoods

Nanoalumina particles and handling procedure are the same as for the case study described in Sections 8.4 and 9.5. For this case study, two different BSCs, two different powder handling enclosures, and the uniquely designed air-curtain hood were evaluated. These hood types are all described earlier in this chapter, and relevant specifications and operating conditions of the evaluated hoods are summarized in Table 9.6.

9.6.2   Measurements

Particles were transferred or poured in each hood. Measurements were taken at the researcher's breathing zone and the source upstream and downstream side locations. Particle release measurements and equipment used for this study are discussed in Sections 8.4 and 9.5. A fog generator and a green-light laser producing a laser-light sheet as described in Section 9.5 were used for visual tests of airflow patterns. Photos of each hood type are shown in Figure 9.28.

9.6.3   Results

The particle releases to worker's breathing zone during particle handling in these nontypical hoods were found to be minimal comparing to the typical hoods discussed in Section 9.5. The magnitude of increased concentration varied by hood type and handling task; however, most measurements were within several hundreds of

(a) (b)

(c)

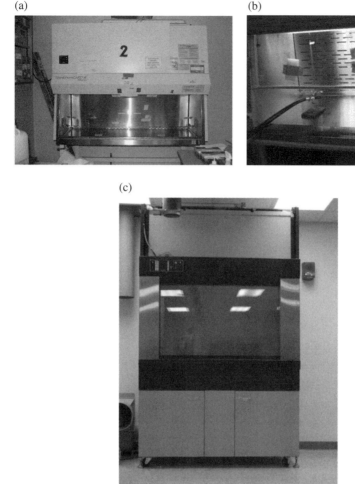

FIGURE 9.28 Photos of hood types referred to in second case study: (a) biological safety cabinet; (b) powder handling enclosure; and (c) air curtain hood. Photos by S. Tsai.

particles per cubic centimeter and the maximum release concentration was about 2000 particles/cm^3. The measured increase in concentration at the worker's breathing zone location are shown in the following figures. For these five hoods, results are presented in two groups based on the nature of the hood design, that is, the BSC and the powder handling enclosures which feature airflow across the sash opening are considered as a group, and the air-curtain hood which features downward airflow at the sash opening is considered individually. Figure 9.29 shows the increased concentrations of the four hoods with airflow across the sash opening; here, the tested conditions for the hoods shown are normal (recommended) operating conditions, which are half-open sash for biosafety cabinet 1, fixed half-open sash for biosafety

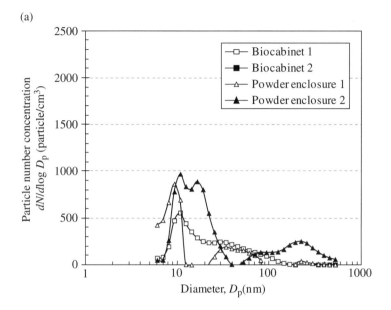

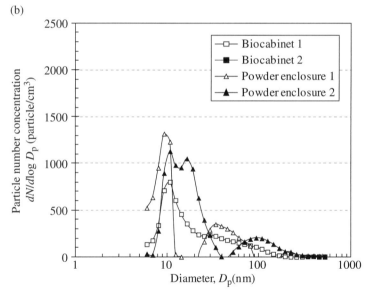

FIGURE 9.29   Particle concentration increase in the breathing zone location during handling 100 g nanoalumina particles in two *biological safety cabinets* and two *powder handling enclosures*: (a) transferring and (b) pouring. Reprinted with permission from Tsai (2013). Copyright 2013 *Journal of Nanoparticle Research*.

cabinet 2, face velocity of 0.38 m/s (75 ft/min) for powder enclosure 1 and face velocity of 0.37 m/s (73 ft/min) for powder enclosure 2. The average increases are mostly less than 1000 particles/cm³, and biosafety cabinet 2 did not induce an elevated concentration during transferring task and had a concentration approaching the baseline at particle size above 400 nm during pouring task. Test results of biosafety cabinet 1, powder enclosure 1, and powder enclosure 2 are within a similar range. Such exposures, when compared to the typical hoods discussed in the previous case study, are acceptable and the hoods were found to be highly efficient in containing nanoparticles within the hoods.

However, can we look more closely at how different or similar the exposure data are within this group? Are they associated with each other? Such association analysis can help us interpret the performance when using these hoods and enclosures. Statistical analysis is the tool to help us to statistically distinguish the results and interpret the similarity and difference of results using systematic analysis. Statistical methods, in particular the Pearson correlation analysis, as described in Section 8.4.1, is used to interpret results for these four hoods. The Pearson correlation coefficient data for the two biosafety cabinets and the two powder-handling enclosures are given in Table 9.7, and the data are plotted in a radar diagram in Figure 9.30. The correlation coefficient curve of biosafety cabinet 2 (Fig. 9.30) to other three hoods shows the association between them. Since all three other hoods have negative correlation coefficients to biosafety cabinet 2, this curve appears to be different and leaning toward the right side, which is opposite to the other three hoods' coefficients that all appear leaning toward the left side. The strong negative association of the biosafety cabinet 2 to other three hoods is obvious, and the strong positive associations of the biosafety cabinet 1 and two powder enclosures are seen with positive coefficients close to 1. Therefore, we can be confident to conclude that the biosafety cabinet 2 provides outstanding performance to effectively contain nanoparticle during particle handling tasks in the hood. However, what feature does the biosafety cabinet 2 have to contribute the distinguish performance?

How was the performance of the air-curtain hood which does not have an airflow across the sash opening? Test results of the air-curtain hood showed that exposure was mostly below 1000 particles/cm³ for both the transfer and pouring tasks as seen

TABLE 9.7   Data of Pearson Correlation Coefficient of Breathing Zone Concentration increases At Four Hoods

	Bio cabinet 1	Bio cabinet 2	Powder enclosure 1	Powder enclosure 2
Bio cabinet 1	1	−0.871	0.487	0.755
Bio cabinet 2	−0.871	1	−0.169	−0.561
Powder enclosure 1	0.487	−0.169	1	0.153
Powder enclosure 2	0.755	−0.561	0.153	1

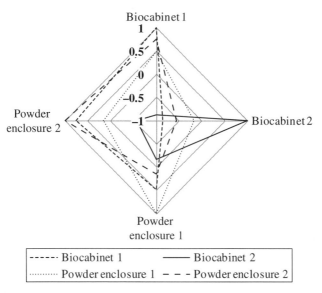

FIGURE 9.30 Correlation diagram of breathing zone concentration increases at four hoods. Reprinted with permission from Tsai (2013). Copyright 2013 *Journal of Nanoparticle Research*.

in Figure 9.31a and b. These results are within the same range as the previously discussed exposure data for the biosafety cabinets and enclosures.

The air-curtain hood has a movable sash and was tested at three sash openings to compare it with typical hoods. The release was barely affected by the sash opening height, which is a high level of stability compared to the typical hoods discussed in Section 9.5. Particles escaping from this hood were always detected at a low level, no matter the location of the sash. However, the more energetic pouring task resulted in a slightly higher release than transferring task.

In addition to the airflow across sash opening, a BSC typically features front exhaust adjacent to the doorsill as seen in Figure 9.32a. The airflow pattern can be observed using fog and green laser shown in Figure 9.32b; the airflow toward to the bottom front is drawn into the front exhaust. The downflow airflow velocity for bio-safety cabinet 1 was 0.28 m/s (55 ft/min) and the velocity for biosafety cabinet 2 was 0.66 m/s (130 ft/min). Biosafety cabinet 2 with a downflow velocity twice as high as biosafety cabinet 1 enhances the drag of contaminated air toward the front exhaust and reduces the chance of nanoparticles escaping the sash opening. This design feature of biosafety cabinet 2 contributes to its outstanding performance for containing airborne nanoparticles in the hood.

Biosafety cabinet 1 and the two powder handling enclosures, which are designed specifically for powder handling (especially for nanomaterials), performed at a comparable level. What features do they have to provide improved performance compared to the typical hoods discussed in Section 9.5? Table 9.6 shows the face velocities used for the biosafety cabinets and powder handling enclosures; in general

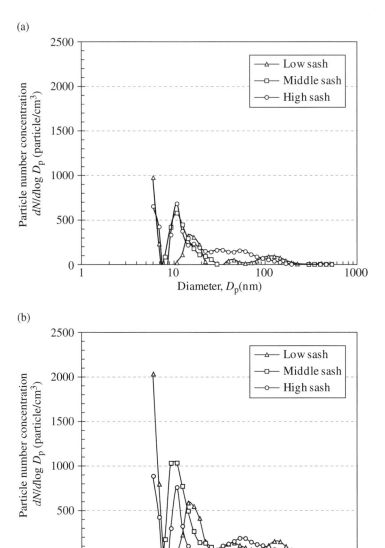

FIGURE 9.31   Particle concentration increase in the breathing zone location during handling 100 g nanoalumina particles in *air-curtain hood*: (a) transferring and (b) pouring.

they are greatly reduced from the face velocity of 0.5 m/s (100 ft/min) commonly recommended for typical hoods. Reduced airflow velocity across sash opening reduces eddies formed in the wake region when the worker is present in front of the sash opening. This is an important feature, but not the only feature to contribute to the improved performance; other features such as the modifications or new designs of the baffle, doorsill, sash, hood ceiling, and airflow distribution across the sash

(a)   (b)

(c)   (d)

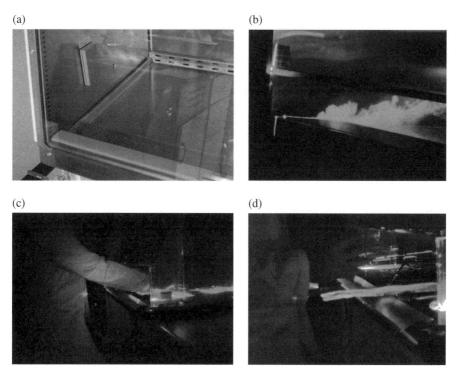

FIGURE 9.32   Photos of biological safety cabinet and powder handling enclosure: features and hand motion: (a) front exhaust slot of biological safety cabinet; (b) fog air flow pattern at front exhaust slot of biological safety cabinet; (c) static hands in the powder enclosure; and (d) hand motion out pulling air flow toward the worker. Photos by S. Tsai.

opening, and so on contribute to the ability to contain contaminants in the hood. However, what is the key issue to cause the (small) releases that were measured?

When a worker handles nanoparticles in the hood, the arms and hands motion during handling generate or enhance turbulence surrounding the arms and hands. Such turbulence will interact with the reverse turbulence in the wake region (when it exists) and nanometer-sized contaminants can be carried out to worker's front vicinity and breathing zone. The fog and laser test can be used to visibly observe the airflow patterns; results are presented in Figure 9.32c and d. The fog was generated continuously inside the hood for these tests. When the worker's hands and arms are positioned steadily in the hood as seen in Figure 9.32c, fog is not observed outside the sash opening (within the wake region); in other words, it is likely that few or no particles are escaping the hood. However, when the worker pulls her arms and hands out, this motion carries fog out to the wake region as seen in Figure 9.32d; the bright light reflection underneath the lab coat sleeve covered arm is the fog being pulled out of the hood.

Hand motion can significantly change the airflow pattern surrounding the moving hand and arm. How about the hand motion effect for the air-curtain hood?

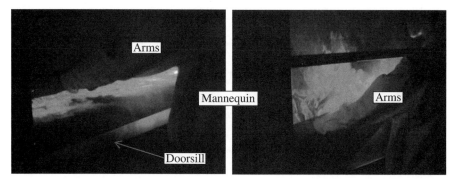

FIGURE 9.33   Features of air flow pattern of air-curtain hood at middle sash: (a) horizontal view and (b) vertical view. Photos by S. Tsai.

The air-curtain hood (Huang et al., 2007) features air exiting from the ceiling top grid and flowing downward through the hood toward the front exhaust right behind the doorsill as seen in Figure 9.33a and b. Because there is no airflow across the sash opening, a wake region does not exist during the use of air-curtain hood. For the fog test, when the worker pulls her arms out of the air-curtain hood, fog could be carried out, but was quickly drawn downward to the front exhaust. The absence of turbulence (wake region) between worker's front trunk and the sash opening eliminates the transport of contaminates when escaping, and the downward air-curtain carries contaminants which are pulled from both sides of the sash opening toward the front exhaust slot. For contaminants such as nanoparticles, which will be carried by the air stream, airflow pattern and the elimination of turbulence play the important role to control exposure.[4]

In summary, the special feature for these high performance hoods that was observed to improve performance is the strong front exhaust of biosafety cabinet 2 and the air-curtain hood, which also introduces a different airflow pattern compared to the typical chemical hoods. A BSC is designed for handling biological samples to avoid contamination and not designed for handling chemicals. However, its feature of controlling biological contamination benefits the containment of airborne nanoparticles.

In terms of using typical fume hoods, when more and more nanomaterials are employed in synthesis or reaction with chemical solvents, what type of hood could give the maximum containment and protection to workers?

9.6.4   The Challenge and Brainstorming

Leaning close to his computer monitor, Steven was searching hard on the web, and then he sat back in the silence, looked at the bloom of maple trees outside his office window, and pondered "what hood I should purchase for the new applications in the

---

[4]Authors' finding: the tested air-curtain hood was a prototype hood provided for laboratory tests, the results of this hood could represent the worst case operation. The commercial model features all specifications required for standard laboratory fume hood and the performance is optimum.

R&D department?" The boss is expecting the proposal in one week, and Steven still has no clue to the best approach.

One of the company's production lines will launch a new nanotechnology application, using nanometer-sized materials instead of the micrometer-sized ones now in use, in order to develop new product lines. The R&D department has an immediate need to handle various types of nanomaterials with solvents, and soon the nanomaterials will be used in the mass production line. In order to handle the new nanomaterials, the R&D facility is undergoing a complete renovation. A handful of manufacturer's catalogs about LEV systems, fume hoods, and so on, are piled up on Steven's desk. "These hoods all look similar to me," Steven thought. He thought that purchasing hoods for the R&D department is the first need, followed by an exhaust system for the production facility. In addition to performing a high level, the control system will have to be durable to function for another decade. How should Steven propose for this project?

9.6.5 Study Questions

1. While we know that high-performance hoods could reduce exposure to minimal levels, how about using glove boxes? What are the benefits or issues of using glove boxes compared to powder handling enclosures?

2. Should Steven recommend the purchase of a biosafety cabinet for the R&D department, even though the company uses no biological materials?

3. Where will Steven find the information he needs to obtain a properly designed control system for the production facility? Does he need to recommend the hiring of a specialist, or can he buy off-the-shelf equipment?

4. Is a powder handling enclosure suitable for handling nanopowders suspended in solvents? If not, what type of hood should Steven consider?

5. Once the new equipment is installed, how can Steven ensure that it is performing properly and protecting the company's workers from exposure to ENPs?

REFERENCES

Altemose BA, Flynn MR, Sprankle J. Application of a tracer gas challenge with a human subject to investigate factors affecting the performance of laboratory fume hoods. Am Ind Hyg Assoc J 1998;59:321–327.

American Conference of Governmental Industrial Hygienists [ACGIH]. Industrial ventilation: a manual of recommended practice. ACGIH; Lansing, MI; 1966.

American Conference of Governmental Industrial Hygienists [ACGIH]. Industrial ventilation: a manual of recommended practice. ACGIH; Lansing, MI; 1980.

American Conference of Governmental Industrial Hygienists [ACGIH]. Industrial ventilation: a manual of recommended practice. ACGIH; Lansing, MI; 1982.

American Conference of Governmental Industrial Hygienists [ACGIH]. Industrial ventilation: a manual of recommended practice. ACGIH; Cincinnati, OH; 1995.

American Conference of Governmental Industrial Hygienists [ACGIH]. Industrial ventilation: a manual of recommended practice for design. ACGIH; Cincinnati, OH; 2007.

ANSI/AIHA. American national standard—laboratory ventilation. American Industrial Hygiene Association Z9.5-2003; 2002.

Burgess WA, Ellenbecker MJ, Treitman RD. *Ventilation for Control of the Work Environment.* New York: Wiley-Interscience; 2004.

Caplan KJ, Knutson GW. Laboratory fume hoods: a performance test. ASHRAE Trans 1978;84:511–521.

DiBerardinis LJ, Baum JS, First MW, Gatwood GT, Groden EF, Seth AK. *Guidelines for Laboratory Design: Health and Safety Considerations.* New York: Wiley Interscience; 1992.

Ellenbecker MJ, Tsai S. Engineered nanoparticles: safer substitutes for toxic materials, or a new hazard? J Clean Prod 2011;19:483–487.

Ellenbecker MJ, Tsai S, Isaacs J. Interim best practices for working with nanoparticles. Lowell (MA): University of Massachusetts Lowell, Center for High-rate Nanomanufacturing; 2008.

European Commission for Standardization. Standard providing the minimum requirements for particle filters. EN 143. Brussels: European Commission for Standardization; 2006.

European Commission for Standardization. Standard providing the minimum requirements for filtering facepieces for protection against particles. EN 149. Brussels: European Commission for Standardization; 2010.

Flynn MR, Ljungqvist B. A review of wake effects on worker exposure. Ann Occup Hyg 1995;39:211–221.

Goodfellow HD, Bender M. Design considerations for fume hoods for process plants. Am Ind Hyg Assoc J 1980;41:473–484.

Goodfellow H, Tahti E. *Industrial Ventilation Design Guidebook.* San Diego: Academic Press; 2001. p 1519.

Harvard University. *Harvard University Center for Nanoscale Systems Safety Manual.* Cambridge (MA): Harvard University; 2011.

Hinds WC. *Aerosol Technology—Properties, Behavior, and Measurement of Airborne Particles.* New York: Wiley-Interscience; 1999.

Hodson L, Crawford C. Guidance for preparation of good material safety data sheets (MSDS) for engineered nanoparticles. Presented at the American Industrial Hygiene Conference and Exposition; Toronto (ON); 2009.

Huang RF, Wu YD, Chen HD, Chen CC, Chen CW, Chang CP, Shih TS. Development and evaluation of an air-curtain fume cabinet with considerations of its aerodynamics. Ann Occup Hyg 2007;51:189–206.

Jasper WJ, Mohan A, Hinestroza J, Barker R. Degradation processes in corona-charged electret filter-media with exposure to ethyl benzene. J Eng Fiber Fabr 2007;2:1–6.

Johnson AE, Fletcher B. The effect of operating conditions on fume cupboard containment. Safety Sci 1996;24:51–60.

Kim TH, Flynn MR. Airflow pattern around a worker in a uniform freestream. Am Ind Hyg Assoc J 1991;52:287–296.

Larese Filone F, D'Agostina F, Croserab M, Adamib G, Renzic N, Bovenzia M, Maina G. Human skin penetration of silver nanoparticles through intact and damaged skin. Toxicology 2009;255:33–37.

Matus KJM, Hutchinson JE, Peoples R, Rung S, Tanguay RL. *Green Nanotechnology Challenges and Opportunities.* Washington (DC): American Chemical Society Green Chemistry Institute; 2011.

Maynard A, Baron P, Foley M, Shvedova A, Kisin E, Castranova V. Exposure to carbon nanotube material: aerosol release during the handling of unrefined single walled carbon nanotube material. J Toxicol Environ Health A 2004;67:87–107.

NIOSH. Guidance for filtration and air cleaning systems to protect building environments from airborne chemical, biological, or radiological attacks. DHHS (NIOSH) Publication No. 2003–136. Washington (DC): U.S. National Institute for Occupational Safety and Health; 2003.

Occupational Safety and Health Administration [OSHA]. General Industry: OSHA Safety and Health Standards. OSHA 29 CFR 1910:134. Washington (DC): OSHA; 1992.

Occupational Safety and Health Administration [OSHA]. Respiratory Protection. OSHA 29 CFR 1919.134. Washington (DC): OSHA; 1998.

Oklahoma University. (undated) Nanoparticle Handling Guidelines. Norman (OK): Oklahoma University.

Pathanjali C, Rahman MM. Study of flow patterns in fume hood enclosures. Energy Conversion Engineering Conference, 1996 IECEC 96 Proceedings of the 31st Intersociety; Washington, DC; 1996.

Popendorf W. *Industrial Hygiene Control of Airborne Chemical Hazards*. Boca Raton (FL): Taylor & Francis; 2006.

Rengasamy S, Eimer BC. Nanoparticle filtration performance of NIOSH-certified particulate air-purifying filtering facepiece respirators: evaluation by light scattering photometric and particle number-based test methods. J Occup Environ Hyg 2012;9:99–109.

Rengasamy S, King WP, Eimer BC, Shaffer RE. Filtration performance of NIOSH-approved N95 and P100 filtering facepiece respirators against 4 to 30 nanometer-size nanoparticles. J Occup Environ Hyg 2008;5:556–564.

Rengasamy S, Miller A, Eimer BC. Evaluation of the filtration performance of NIOSH-approved N95 filtering face piece respirators by photometric and number-based test methods. J Occup Environ Hyg 2011;8:23–30.

Rouse JG, Yang J, Ryman-Rasmussen JP, Barron AR, Monteiro-Riviere NA. Effects of mechanical flexion on the penetration of fullerene amino acid-derivatized peptide nanoparticles through skin. Nano Lett 2007;7:155–160.

Ryman-Rasmussen JP, Riviere JE, Monteiro-Riviere NA. Penetration of intact skin by quantum dots with diverse physicochemical properties. Toxicol Sci 2006;91:159–165.

Saunders GT. *Laboratory Fume Hoods: A User's Manual*. New York: Wiley-Interscience; 1993.

Schulte HF, Hyatt EC, Jordan HS, Mitchell RN. Evaluation of laboratory fume hoods. Am Ind Hyg Assoc J 1954;15:195–202.

Taylor S. *Fume Hood Study*. Medford (MA): Tufts University; 2004.

Tinkle SS, Antonini JM, Rich BA, Roberts JR, Salmen R, Depree K, Adkins EJ. Skin as a route of exposure and sensitization in chronic beryllium disease. Environ Health Perspect 2003;111:1202–1208.

Tsai SJ. Potential inhalation exposure and containment efficiency when using hoods for handling nanoparticles. J Nanopart Res 2013;15:1880.

Tsai SJ, Ada E, Isaacs J, Ellenbecker MJ. Airborne nanoparticle exposures associated with the manual handling of nanoalumina and nanosilver in fume hoods. J Nanopart Res 2009a;11:147–161.

Tsai SJ, Hofmann M, Hallock M, Ada E, Kong J, Ellenbecker MJ. Characterization and evaluation of nanoparticle release during the synthesis of single-walled and multi-walled carbon nanotubes by chemical vapor deposition. Environ Sci Technol 2009b;43:6017–6023.

Tsai SJ, Huang RF, Ellenbecker MJ. Airborne nanoparticle exposures while using constant-flow, constant-velocity, and air-curtain-isolated fume hoods. Ann Occup Hyg 2010;54:78–87.

Tschudi W, Bell GC, Sartor D. Side-by-side fume hood testing: human-as-Mannequin report—comparison of a conventional and a Berkeley fume hood. Berkeley (CA): U.S. Department of Energy, Lawrence Berkeley National Laboratory; 2004.

Tseng L, Huang RF, Chen CC, Chang CP. Correlation between airflow patterns and performance of a laboratory fume hood. J Occup Environ Hyg 2006;3:694–706.

University of Pennsylvania. Nanoparticle handling fact sheet. Philadelphia (PA): University of Pennsylvania; 2008.

U.S. Department of Energy [DOE]. Nanoscale Science Research Center: approach to nanomaterial ES&H, Rev.3a. U.S. Washington (DC): DOE; 2008.

U.S. National Institute for Occupational Safety and Health [NIOSH]. Guide to Industrial Respiratory Protection. DHHS (NIOSH) Publication No. 87-116. Cincinnati (OH): NIOSH; 1987.

U.S. National Institute for Occupational Safety and Health [NIOSH]. Approval of respiratory protection devices. 42 CFR 84. Washington (DC): NIOSH; 1995.

U.S. National Institute for Occupational Safety and Health [NIOSH]. General Safe Practices for Working with Engineered Nanomaterials in Research Laboratories. DHHS (NIOSH) Publication No. 2012-147. Cincinnati (OH): NIOSH; 2012.

10

CONTROL OF ENVIRONMENTAL EXPOSURES

As indicated in Chapter 5, as of this writing, much less attention has been paid to environmental exposures than to those in occupational settings. Not surprisingly, then, there is a similar lack of attention to the *control* of environmental exposures. This chapter will summarize what is known about such controls as they specifically relate to ENPs, but it will also discuss how effective such controls are *likely* to be and consider areas where future research is needed.

10.1 CONTROL OF AIR EMISSIONS

A great deal of information exists, both practical and theoretical, concerning the control of particulate air emissions from industrial facilities. What is lacking is the application of this information to particles in the nanometer size range. This section will begin by reviewing the basic categories of devices used for particle control and then discussing their application to engineered nanoparticles (ENPs). It is important to recognize that the devices described here typically are incorporated into a local exhaust system, which serves to capture the particles; such systems were discussed in detail in Chapter 9.

Exposure Assessment and Safety Considerations for Working with Engineered Nanoparticles,
First Edition. Michael J. Ellenbecker and Candace Su-Jung Tsai.
© 2015 John Wiley & Sons, Inc. Published 2015 by John Wiley & Sons, Inc.

10.1.1 Factors Affecting Air Cleaner Performance

The two most important factors for characterizing an air cleaner's performance are its collection efficiency (CE) and pressure drop. Equations (9.2) and (9.3), defining mass and number collection efficiency respectively for a respirator filter, apply more broadly to all air cleaning devices. The pressure drop discussion in Chapter 8 also applies here, except of course air cleaners use a fan or blower to overcome the device pressure drop rather than a worker's lungs.

Finally, there are a range of practical factors that contribute to the utility of a particular device, including its cost, size, energy use (other than pressure drop), maintenance requirements, and so on. Probably the most important of these secondary factors is the ability of the device to be cleaned of the collected dust, since this extends the useful life of the device, a big factor in reducing its cost.

10.1.2 Categories of Air Cleaning Devices

Aerosol particles can be removed from the air by four basic mechanisms: inertial collection, methods utilizing liquids as the collection medium, electrostatic precipitation, and filtration. As we shall see, most of these methods have significant problems when it comes to collecting nanoparticles.

10.1.2.1 Inertial Collection   Inertial collectors, as their name implies, rely on the inertial properties of the aerosol to remove them from the gas stream. The simplest is the gravity settling chamber (Fig. 10.1), which essentially is an expansion of the cross-section of the duct carrying the particles. This expansion slows down the transport velocity of the gas, allowing particles time to fall into the hopper by gravity settling. However, as shown in Table 2.1, nanoparticles essentially have *no* settling velocity, meaning that this device type is not appropriate for nanoparticle collection.

The other primary type of inertial collector is the cyclone (Fig. 10.2); here, particle-laden air enters tangentially, makes several circles, and exits through the central tube. Particles are thrown toward the wall of the cyclone due to centrifugal force; in a properly designed cyclone, the force can be many times the force of

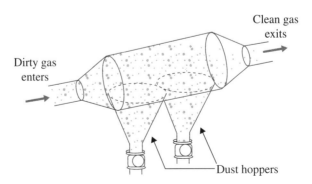

FIGURE 10.1   Gravity settling chamber. Diagram courtesy of U.S. EPA.

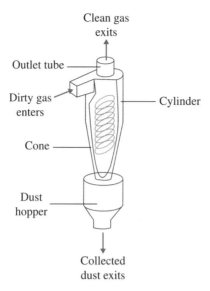

Clean gas
exits

Outlet tube

Dirty gas
enters

Cone

Cylinder

Dust
hopper

Collected
dust exits

FIGURE 10.2   Cyclone. Diagram courtesy of U.S. EPA.

gravity ("g's"), meaning that cyclones can efficiently collect smaller particles than a gravity settling chamber. This efficient collection, however, does not extend to the nanoparticle size range.

As discussed in Chapter 7, it is possible to collect nanoparticles inertially through the use of specially designed low-pressure impactors. Such devices work for sample collection but are not appropriate for large-scale air cleaning. It is safe to conclude that no inertial device will work effectively for collecting nanoparticles.

10.1.2.2 Collection in Liquids   There are several device types that remove particles from a gas stream by contact with a liquid; the most common is the venturi scrubber (Fig. 10.3), a very simple device both in construction and in operation. Exhaust gases are accelerated to high velocity through the throat of the venturi, where water is introduced. The high-velocity gas shears the water stream into droplets and the high relative velocity between the just-formed droplets and the particles moving at the high gas velocity cause particles to be collected on the droplets, primarily by inertial impaction. The droplets, which are much larger than the collected particles, are subsequently removed from the gas stream in a cyclone or entrainment (gravity) separator. A major disadvantage of any particle scrubber, of course, is that even if it collects particles with high efficiency it essentially trades an air pollution problem for a water pollution one, since the particles are now suspended in the water exiting the scrubber and must subsequently be removed.

As discussed in Chapter 8, inertial impaction is strongly favored by larger particles and decreases rapidly for nanometer-sized particles. Nonetheless, the high relative velocities between particles and droplets in venturi scrubbers make it possible

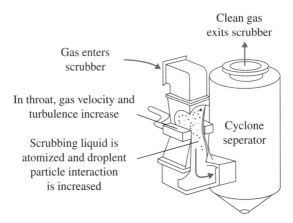

FIGURE 10.3   Venturi scrubber. Diagram courtesy of U.S. EPA.

to efficiently collect even submicrometer particles. The collection of submicrometer particles by a venturi scrubber was evaluated by Tsai and colleagues (2005), who found that the CE was very high and consistent with theory for particles larger than 100 nm, but decreased rapidly for smaller particles, even though the theoretical collection was still high (see Fig. 10.4). In their scrubber, a fine water droplet mist was introduced to create a super-saturated condition, causing nucleation on fine particles and subsequent growth to larger sizes that can be collected with high efficiency. Apparently this method worked as predicted theoretically for particles as small as 100 nm, but did not work for smaller particles.

10.1.2.3 Electrostatic Precipitation   Electrostatic precipitators (ESPs) constitute one of the two most commonly used large-scale particle removal devices used in the industries (the other being fabric filters, discussed in the following). A typical industrial ESP is shown in Figure 10.5. This device removes particles from an exhaust stream in three steps. Particles are first imparted an electrical charge by passing them through cloud of ions produced by a corona discharge. Next, the charged particles are passed through an electric field and attracted toward and collected on grounded plates by the electrostatic force between the field and the excess charges on the particle. Finally, the collected particles are removed from the collecting plates to a hopper.

A detailed discussion of ESP design principles is beyond the scope of this text. One aspect of their design that is directly relevant to their performance in collecting nanoparticles deserves some discussion, however. An ESP relies on particles collecting a significant number of charges (in commercial ESPs, electrons) from the corona discharge in order to develop a strong enough electrostatic force to move the particles to the collection plates before the gas exits the device. Since the electrons collected on a particle repulse one another, they tend to distribute themselves equally on the particle surface. A maximum charge level is eventually reached when the collected charges are sufficient to repel any additional electrons from the particle. This

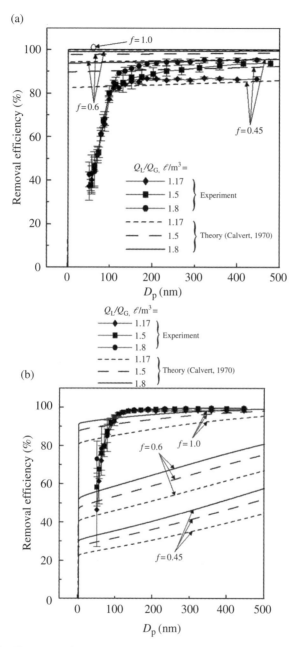

FIGURE 10.4 The comparison of the theoretical results of particle removal efficiency with experimental data with nucleation by water mist. (a) One percent SiH4 of 0.5 l/min and (b) 1% SiH4 of 1 l/min (Tsai et al., 2005). Used by permission, Taylor & Francis.

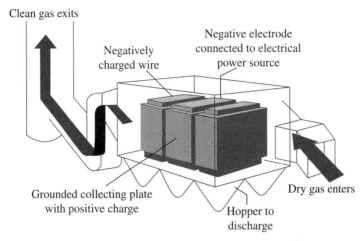

Clean gas exits

Negatively charged wire

Negative electrode connected to electrical power source

Grounded collecting plate with positive charge

Dry gas enters

Hopper to discharge

FIGURE 10.5   Electrostatic precipitator. Diagram courtesy of U.S. EPA.

saturation charge is a function of the particle surface area, and consequently, since surface area decreases as the square of particle diameter, decreases rapidly for particle in the nanometer size range. Whereas the saturation charge on a 1 μm diameter particle is about 100 electrons under typical ESP conditions, theoretical (Hinds, 1999; Lin et al., 2012) and limited experimental (Lin et al., 2012; Mohr et al., 2007) data indicate that particles 10 nm in diameter and smaller are likely to have *zero* excess charges and thus not affected by the precipitator electric field. Pui et al. (1988) investigated the unipolar charging of nanoparticles as part of their design of the differential mobility analyzer (see Section 7.2.3). They found that the fraction of 5 nm particles receiving a charge varied from about 5 to 17%, depending on the value of the product of ion concentration and exposure time. For 50 nm particles, the range was 80–99%.

Lin et al. review the limited published data indicating that, due to the lack of particle charging, CE decreases rapidly when the particle diameter falls below about 20 nm. The low level of particle charging at the small end of the ENP size range leads to the likely conclusion that ESPs will not prove to be effective devices for collection of the smallest nanoparticle.

10.1.2.4   Filtration   The basic properties of respirator filters, described in Chapter 9, apply also to air cleaning filters. There are many practical differences, of course, the primary ones including the volume of gas to be filtered, the gas temperature, and the desirability of using cleanable filtration systems to reduce operating cost. In any case, industrial air filters rely on the same particle collection mechanisms as respirator and air sampling filters, that is, inertial impaction, interception, diffusion, electrostatic forces, and gravity settling. In addition, many air-cleaning filters have a particle CE curve that is very similar to that shown in Figure 9.12.

The two types of industrial filters in common use are the so-called high efficiency particulate air (HEPA) filters and fabric filters.

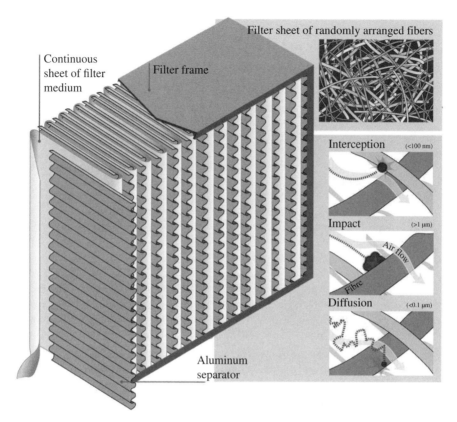

FIGURE 10.6 High-efficiency particulate air (HEPA) filter. Image used by permission, Wikimedia/LadyofHats.

HEPA Filters HEPA filters for air cleaning (Figure 10.6) are similar to respirator HEPA filter cartridges, discussed in Chapter 8, but are constructed on a larger scale and are designed for use whenever extremely high CE is required. The filter material and principles of particle collection are the same in an industrial HEPA filter as in one used for respirators. The important advantages of HEPA filters described in Chapter 8 for respirator filters come at a significant price however, when used for industrial air cleaning. The filters are not cleanable, and the close spacing between the filter folds makes the dust-holding capacity of these filters very low. The combination of lack of cleanability with low dust capacity limits the use of these filters to gas streams with very low dust loading.

Fabric Filters Fabric filters (Fig. 10.7) are by far the most common type of filter encountered in the industrial environment and the one most likely to be used in a local exhaust ventilation system. The filter itself consists of pieces of fabric sewn into cylinders or envelopes and mounted in a housing.

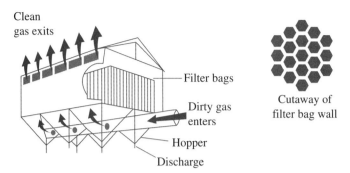

FIGURE 10.7   Fabric filter. Diagram courtesy of U.S. EPA.

During operation, exhaust air is drawn through the fabric by the fan; particles either collect in the fabric itself or in a dust cake on the fabric surface and are thus removed from the exiting exhaust stream.

The principal advantage that fabric filters have over HEPA filters is that they are cleanable. In fact, cleaning is so important that fabric filters are classified by the cleaning method used. The three most common cleaning methods are shaking, reverse air flow, and pulse jet air. Woven fabrics are used in shaker and reverse air systems. These fabrics, when new, have large holes (on the microscopic scale) between the fibers making up the weave. They tend to have very poor particle CE until those holes are bridged over by collected particles. Once bridging occurs, a dust cake is formed on the fabric surface, which removes incoming particles with high efficiency. In shaking, woven fabrics are shaken from the top in order to dislodge agglomerated particles from the fabric surface, which then fall into the hopper. Reverse-air cleaning uses (surprise!) a reverse air flow to dislodge agglomerates from the woven fabric surface. Pulse jet cleaning uses short blasts of compressed air to dislodge agglomerated particles from the fabric.

Pulse jet cleaning requires the use of felted fabrics, rather than woven ones, to withstand the violent motion induced by the pulses. Felted fabrics tend to collect dust throughout the depth of the felt rather than at the surface as a cake; they rely on the classic filtration methods described in Chapter 8 rather than on cake formation. Unlike woven fabrics, new felted fabrics should have a collection efficiency that is similar to that of the old ones.

Once a woven fabric has established a dust cake, it is said to be conditioned, and conditioned fabrics usually have close to 100% CE. This is certainly true if the aerosol being filtered consists of monodisperse spherical particles. Since the air passages through such a dust cake are smaller than the particles that make it up (visualize a jar of marbles), a conditioned fabric removes 100% of such particles. Unfortunately, things are not so simple for real industrial aerosols, which are polydisperse and irregular in shape. To visualize this, consider a new example where we have both marbles and much smaller spheres. Now, the marbles will make a porous structure that will sieve out any new marbles, but the smaller spheres will be able to make their way through the spaces between the marbles.

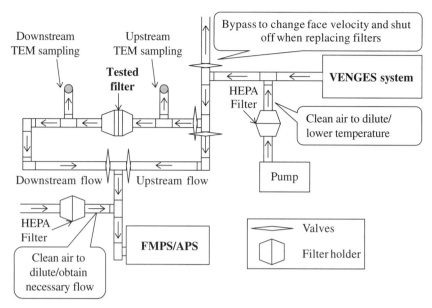

FIGURE 10.8   SEM images of a sample of woven polyester fabric with a Teflon membrane coating. Image courtesy E. Ada, University of Massachusetts Lowell.

Getting back to nanoparticle filtration, consider a real polydisperse aerosol with both micrometer- and nanometer-sized particles. The micrometer-sized particles will establish a porous dust cake, but the nanometer-sized particles may very well be able to penetrate through the cake passages and fail to be collected. The extent to which this actually happens can only be answered through testing.

Unfortunately, there are no studies in the published literature regarding the ability of fabric filters to collect ENPs. Recently, we completed pilot testing of small samples of six typical filter fabrics where we generated ENPs and measured their CE at two different filtration velocities (Tsai et al., 2012). The test setup is shown in Figure 10.8. The tested fabrics included two plain woven fabrics and one polyester felt (PF) fabric, and three additional fabrics where each of the three plain fabrics was coated with either a Teflon© or Goretex© membrane. Since only fabric samples were tested while mounted in a filter holder, only the clean fabric CEs were measured. Test SiO_2 nanoparticles with diameters from 50 to 150 nm (Fig. 10.9) were generated by the Harvard Versatile Engineered Nanomaterial Generation System (VENGES) developed by Demokritou (Demokritou et al., 2010) and passed through the test fabric samples, and the CE as a function of particle size was measured at two different filtration velocities using the fast mobility particle sizer (FMPS—see Section 7.2.4). The results are summarized in Figure 10.10; not surprisingly, the uncoated woven fabrics performed the worst, with CE less than 50% at the lower velocity and less than 30% at the higher velocity. The uncoated PF performed somewhat better (CE = 64 and 42%), but its CE still would not be considered acceptable. The coated fabrics performed much better, with CEs ranging from 87 to 96% at the lower velocity and 51–74% at the higher velocity.

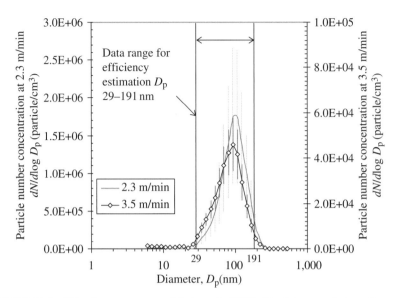

FIGURE 10.9   Filtration test system (Tsai et al. 2012). Used by permission, *Journal of Nanoparticle Research.*

The benefit of the fabric coating is obvious in the data; this is due to the smaller pores in the coating compared to the fabric itself, as shown in the SEM image of one of the fabrics shown in Figure 10.11, which significantly improved the clean fabric performance. There is a penalty to be paid in using a coated fabric, however, that is, the pressure drop across an uncoated fabric typically was about twice that across the uncoated version of the same fabric. Better performance at the lower velocity was expected, since, as discussed in Section 8.4.2, nanoparticles would collect on filter fibers by Brownian motion, and a lower velocity provides a longer residence time as a particle passes by a fiber, increasing the probability of diffusion to the fiber surface. The performance of various fabric filter fabrics is discussed in more detail in the case study in Section 10.6.

Since fabric filters are not 100% efficient in collecting ENPs but are cleanable, and HEPA filters are essentially 100% efficient and not cleanable, a feasible approach for effective control would be a two-stage filtration system, that is, a fabric filter followed by a HEPA filter. The fabric filter would collect almost all of the particles, and the HEPA filter would collect the few that penetrate.

10.2   CONTROL OF WATER EMISSIONS

As discussed in Chapter 5, the rapid increase in the manufacture and use of ENPs greatly increases the likelihood of their migration to water treatment plants, both those located at manufacturing facilities and publicly owned treatment works. Two

(a)

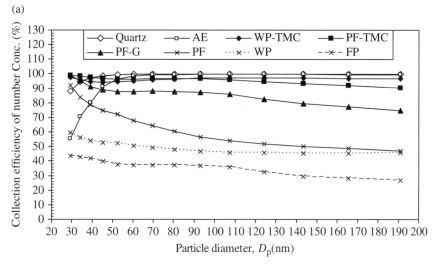

(b)

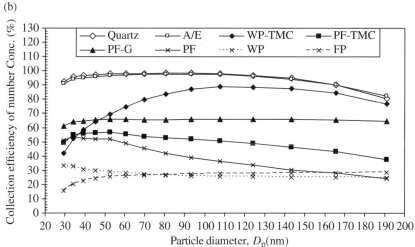

FIGURE 10.10 Particle size distribution and concentration generated by VENGES. Primary *y*-axis: Upstream concentration at 2.3 m/min filtration face velocity; Secondary *y*-axis: Upstream concentration at 3.5 m/min filtration face velocity (Tsai et al., 2012). Used by permission, *Journal of Nanoparticle Research*.

main questions arise when considering the treatment of wastewater containing ENPs, that is, the effectiveness of the treatment plant in removing the ENPs from wastewater and the possible adverse effect that the ENPs may have on the overall performance of the plant.

There are relatively few studies in the literature on the effectiveness of treatment plants in removing ENPs (Limbach et al., 2008). Weisner et al. (2006) summarized the current state of this research in 2006 by the statement "the transport and fate of

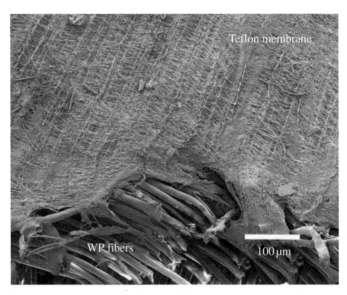

FIGURE 10.11   Particle number collection efficiency versus particle diameter. (a) Six filters tested at 2.3 m/min filtration velocity and (b) six filters tested at 3.5 m/min filtration
velocity. Note: Hollow symbol is air sampling filters, solid symbol is coated air cleaning
filter, X symbol is noncoated air cleaning filter (Tsai et al., 2012). Used by permission,
Journal of Nanoparticle Research.

nanomaterials in aquatic environments has received relatively little attention." In the
same year Reijinders (2006) concluded that "standard wastewater treatment seems to
be poorly suited to the capture of nanoparticles."

Limbach (Limbach et al., 2008) studied the removal of cerium oxide nanoparticles using a scale model sewage treatment system. The particle diameters ranged
from 30 to 50 nm, and dispersions of the particles with a concentration of 100 ppm
by mass $(1.7 \times 10^{12}$ particles/cm^3) were fed continuously to the model sludge
system. During the initial phase of each experiment, as high as 10% of the ceria
particles were found in the waste water exiting the treatment plant, but then the
penetration leveled off in the range of 2–5%. Examination of the sludge found that
the ceria particles tended to agglomerate and to attach themselves preferentially to
the cell walls of bacteria, both being mechanisms that would enhance particle
retention in the sludge. The efficiency of the treatment system was found to be a
function of several variables, including the surfactant used to disperse the nanoparticles in water. In addition, the authors believed that the high nanoparticle
concentration tested favored agglomeration and that treatment may be much less
efficient at the much lower concentrations expected in actual waste streams. The
authors recommended further studies of more dilute particle concentrations and
using actual sewage treatment plants.

Given the toxicity of many nanoparticles, it is possible that the presence of these
particles in water fed to sewage treatment plants may kill some of the bacteria in the

sludge that are essential for proper activity, and thus may decrease the efficiency of the sludge in general. Since nanoscale zinc oxide is now commonly found in sunscreen treatments, it is entirely feasible that it could be washed off the skin and make its way to a sewage treatment plant. One recent study (Mu and Chen, 2011) demonstrated that nanoscale zinc oxide can adversely affect the anaerobic digestion capability of a typical sludge treatment plant.

Another obvious candidate for such adverse effects would be nanosilver, which has been widely incorporated into clothing fabrics in recent years to serve as an antimicrobial agent. Since the nanosilver clearly is designed to kill bacteria, if washing releases nanosilver from the clothing, could the nanoparticles then kill bacteria once they reach the sewage treatment plant? Benn and Westerhoff (2008) investigated the ability of typical washing processes to release nanosilver from commercial sock fabrics. The sock types they tested contained up to 1360 μg Ag/g of fabric, and their model washing released as much as 650 μg Ag into 500 ml of distilled water, with released particle diameters ranging from 10 to 500 nm. Testing of the silver-contaminated water in model sludge demonstrated that the sludge could efficiently remove the silver from the wastewater, but the authors unfortunately did not investigate whether the nanosilver had any adverse effects on the sludge biomass.

This subject has also been investigated by Zhang et al., who studied the effects of purified carbon nanotubes on a model activated sludge process (Goyal et al., 2010; Luongo and Zhang, 2010; Yin et al., 2009). Batch sludge reactors were exposed to both single-walled CNTs (SWCNTs) (Goyal et al., 2010; Yin et al., 2009) and multiwalled CNTs (MWCNTs) (Luongo and Zhang, 2010), and the results were not clear cut. The SWCNTs had a beneficial effect on some reactor factors, such as an increase in the removal efficiency of soluble chemical oxygen demand and an increase in sludge settleability, and had no effect on others. The bacterial community was modified by the SWCNTs, although the effects of such modifications on sludge performance were uncertain. The tests with MWCNTs demonstrated definite adverse effects in the form of respiratory inhibition of the bacterial community.

A porous medium, such as a sand bed, is frequently used to remove particles from a water stream. Colloid removal in a porous medium can occur either by sieving, where the pore diameters are smaller than the particles (similar to the cake filtration described under fabric filtration), or by the classical filtration methods described in Chapter 8. For nanometer-sized particles, it is likely that the pores of most porous media will be too large for sieving, so that classical filtration will apply. Since filtration theory is based on fluid mechanics, it applies equally to gas and liquid filtration. As with air filtration, inertial impaction and interception in water will be much less important than diffusion due to Brownian motion in the nanoparticle size range. In general, particles of a given diameter and density will move more slowly by diffusion and gravity settling in water as compared to air, since water has a much higher density and viscosity than air. According to Baalousha et al. (2009), a typical granular water filter has a minimum collection efficiency at about 1 μm; collection efficiency increases for larger particles due to inertial collection and for smaller particles due to diffusion.

10.3 NANOPARTICLES IN SOLID WASTE

It is clear that many ENPs are now ending up in solid waste, and the amount in this waste stream is likely to increase rapidly. There are several sources for such particles, including:

- contaminated personal protective equipment worn by laboratory and production workers
- ENPs collected by control equipment, such as HEPA and fabric filters
- waste from the cleaning of production equipment
- discarded production runs
- products containing ENPs that are discarded at their end of life.

Once ENPs enter the solid waste stream, the concern is that they will reenter the air and/or water, causing further opportunities for exposure to workers and the general population. Such concerns are not unique to ENPs, of course; environmental health professionals have long had worries about toxic substances in solid waste. In the United States, tragic examples of contamination such as Love Canal led to the passage of the Toxic Substance Control Act (TSCA) and the Resource Conservation and Recovery Act (RCRA) in 1976 and the Comprehensive Environmental Response, Compensation, and Liability Act (CERCLA) (also known as Superfund) in 1980. All of these laws attempt to regulate the contamination of solid waste by toxic chemicals; the problem we now have, as discussed in Section 11.2, is that with a few exceptions they do not apply directly to ENPs. As a result, even though some organizations recommend or require that waste containing ENPS be considered *hazardous* waste and be handled accordingly (Ellenbecker et al., 2008; Hallock et al., 2009; University of Pennsylvania, 2008; VCU, 2007), at this time, there is no *requirement* that they do so.

There is little information in the published literature concerning the leachability of nanoparticles from landfills or solid waste. As of this writing, the environmental laboratory at the US Army Engineer Research and Development Center was conducting research on this topic. In a paper titled "Soil Leachability of Silver Nanoparticles using a Column Method" presented at the Seventh International Symposium on Recent Advances in Environmental Health Research in 2010, they reported experimental results indicating that both nanometer- and micrometer-sized silver particles leached out of spiked soil samples using a slightly acidic leachate solution, but that the nanosized Ag leached in lesser amounts than the micrometer-sized Ag.

EPA is currently funding research on the leaching of nanoparticles from solid waste. For example, one project at the University of Missouri, Columbia,[1] titled "Bioavailability of Metallic Nanoparticles and Heavy Metals in Landfills" has as its objective:

> The objectives of this research are to determine bioavailability of nanoparticles and heavy metal species in bioreactor landfills as compared to traditional municipal solid waste landfills; and to elucidate the mechanisms governing bioavailability as well as the

[1] http://cfpub.epa.gov/ncer_abstracts/index.cfm/fuseaction/display.abstractDetail/abstract/8979

mode of antimicrobial action by nanoparticles. In order to accomplish these objectives, three hypotheses will be tested: 1) nanoparticles that can leach into the water from landfill runoff are bioavailable; 2) nanoparticles enhance heavy metal bioavailability and leachate toxicity; and 3) the bioavailability of nanoparticles and heavy metals is higher in bioreactor landfills than in traditional landfills.

In addition, the Environmental Research and Education is funding a proposal at the University of Central Florida[2] titled "Fate of Nanoparticles in Municipal Solid Waste Landfills" with the following description:

The goal of this proposed research is to evaluate the fate of anthropogenic inorganic engineered nanoparticles (ENPs) placed in municipal solid waste landfills. We will focus on interactions between ENPs and leachate constituents based on the assumption that ENPs will mobilize in the liquid phase at some point in time. The specific objectives are to evaluate how the presence of ENPs will affect biological landfill processes, how the presence of leachate constituents (e.g. organic matter, ammonia, and pH) will affect ENP characteristics and fate, and how the presence of ENPs will affect leachate treatment.

10.4   CONTROL OF EXPOSURES THROUGHOUT A PRODUCT'S LIFE CYCLE

To this point, this chapter has focused on controlling the environmental impact of ENPs during the manufacturing process, during either the manufacturing of the particles themselves or the incorporation of the particles into final products. However, as discussed in Section 6.5.2, attention must be paid to the environmental impacts of a product throughout its life cycle. The one phase of a product's life cycle that has not been discussed in terms of control is product use. In general, exposure control during product use must be addressed during the product design phase by using inherently safe design, since it is difficult or impossible to affect exposure once the product is in the hands of the consumer.

Some categories of nanoparticle-enabled consumer products now on the market have the nanoparticles tightly imbedded in a matrix; these products are inherently safe when used. Examples include nanoparticle composites used in athletic equipment, car bumpers, and so on, and surface coatings. Unfortunately, other product categories inherently present a risk for consumer exposure. Any product designed for use on the skin, such as sunscreens and facial creams, presents a risk for dermal penetration (see Section 7.3). Products incorporating ENPs that are meant to be sprayed present a significant inhalation hazard (Chen et al., 2010; Nazarenko et al., 2011), as do cosmetic powders (Nazarenko et al., 2012). Careful consideration must be given to the advisability of marketing any products that fall into these categories, since exposures seem certain.

---

[2] https://argis.research.ucf.edu/index.cfm?fuseaction=detail.view_proposal_detail&rec_id=1049526&rec_type=research

Finally, there are a range of products that fall between the two extremes described earlier. For example, does nanoenabled clothing, such as socks and underwear which incorporate nanosilver for its antimicrobial properties, present any significant risk of exposure to the wearer, or perhaps the washer of the clothes? Evidence in the literature suggests that, indeed, sweat can release silver from clothing (Kulthong et al., 2010) and washing can release significant fractions of the nanoparticles into the wash water (Geranio et al., 2009).

10.5   UNCERTAINTIES AND NEEDED RESEARCH

As discussed throughout this chapter, there is very little in the air and water pollution control literature that specifically addresses the effectiveness of such devices in the nanometer size range. We have the most certainty about the performance of HEPA filters, which appear to collect ENPs with close to 100% efficiency; even here, however, the tests were performed on HEPA filter respirator cartridges, not the larger filters used for air cleaning. Those tests evaluated filter performance for particles with diameters as small as 4 nm (Rengasamy et al., 2008), so uncertainty remains as to HEPA filter effectiveness for the smallest ENPs. Based on filtration theory, we would expect those particles to be collected with high efficiency, but there is concern that thermal rebound may become important. Thermal rebound essentially is what occurs when a gas molecule strikes a filter element; such molecules have sufficient thermal energy, relative to their mass, to overcome attractive forces at the moment of contact and thus rebound (we are lucky, since if this did not occur, the earth would have no atmosphere!). There is concern that as particle diameters approach 1 nm, their thermal energy relative to their mass may be sufficient to cause thermal rebound; this requires further study.

Fabric filters may offer promise as an effective air pollution control device, especially if used in conjunction with HEPA filters. The case study that follows this section presents results of our preliminary investigation into the performance of various types of fabric filter fabrics in collecting ENPs. This work is encouraging, but must be followed up with pilot-scale testing of fully operational fabric filters.

For reasons discussed in the earlier sections, it is not likely that other categories of air pollution control devices will be effective for removing ENPs. In the case of electrostatic precipitation, this needs to be confirmed experimentally.

10.6   CASE STUDY—FILTRATION CONTROL

Applying engineering controls to airborne ENPs is critical to prevent environmental releases and worker exposure. This study evaluated the effectiveness of two air sampling and six air cleaning fabric filters at collecting ENPs using industrially relevant flame-made ENPs generated using a Versatile Engineered Nanomaterial Generation System (VENGES), recently designed and constructed at Harvard University. VENGES has the ability to generate metal and metal oxide exposure atmospheres while controlling important particle properties such as primary particle size, aerosol size distribution, and agglomeration state. For this study, amorphous SiO_2 ENPs with a 15.4nm primary particle size were generated and diluted with HEPA-filtered air. The aerosol was passed through the filter samples at two different filtration face velocities (2.3 and 3.5 m/min). Particle concentrations as a function of particle size were measured upstream and downstream of the filters using a specially designed filter test system to evaluate filtration efficiency. Real time instruments (FMPS and APS) were used to measure particle concentration for diameters from 5 to 20,000 nm.

The aim of this study was to better understand how to effectively mitigate the environmental and health effects caused by nanoparticles. Engineering controls such as filtration are essential for reducing ENP emission to the environment and minimizing occupational exposures during production or use of these nanomaterials.

Case theme: Experimental study to evaluate various filters for collecting ENPs in workplaces.

Operation: Design a filtration test system to demonstrate the practical use of filters.

Investigation: The actual performance of using various filters and the affecting factors to the performance.

10.6.1 Materials and Process

Evaluated filters include quartz (Whatman QMA quartz fiber, 47mm D, BGI Inc.) and glass fiber (indicated as A/E) (Pall A/E glass fiber, 47mm D, BGI Inc.) air sampling filters and six air cleaning fabric filters,[3] that is, woven polyester (WP), WP with Teflon membrane coating (WP-TMC), PF, PF with Teflon membrane (PF-TMC), PF with Goretex membrane coating (PF-GM), and filament polyester (FP). None of the eight filters have collection efficiency values claimed by the manufacturers for nanometer-sized particles. The tested filter media can be categorized into three groups: (1) the high efficiency quartz and glass fiber (A/E) sampling filters, included in the test as high-efficiency filters for comparison with the fabric filters, (2) membrane-coated fabric filters, and (3) noncoated fabric filters. Nanoparticles were generated using the flame spray pyrolysis (FSP) method (Mädler and Friedlander, 2007) using the Harvard Versatile Engineered Nanomaterial Generation System (VENGES) (Demokritou et al., 2010).

The layout of the specially designed filter test system is shown in Figure 10.9. Filters were installed in a filter holder (F1 closed face filter holder, 47mm D, BGI Inc.). Generated nanoparticles were diluted with HEPA-filtered air and passed through filter samples with areas of 17.3 cm^2 at two filtration face velocities (2.3 m/ min (3.8 cm/s) and 3.5 m/min (5.8 cm/s)). The recommended velocity range for woven-fabric baghouses is 0.3–3 m/min (1–10 ft/min) and for nonwoven baghouses

---

[3] The fabric filter samples were manufactured and donated by Process Systems & Components, Inc., Salt Lake City, UT 84103, USA.

is 1.5–3.7 m/min (5–12 ft/min) (Danielson, 1973); the 2.3 m/min face velocity is thus within the recommended velocity range for baghouses using woven and nonwoven (felted) fabric filters and the 3.5 m/min face velocity is within the range recommended for nonwoven fabric filters.

Rotameters were installed to monitor the airflow and filtration velocity and airflow was adjusted with valves installed inline. All tubing (6.35 mm (0.25 in.)) and connectors (standard compression fittings) were stainless steel. Concentrations were measured upstream and downstream of the filters (see monitoring in the following texts), and aerosol particles were collected for characterization by electron microscopy. Samples were collected for short time periods, so the measured collection efficiencies and pressure drops are those of the clean fabrics; collection efficiency may improve and pressure drop increase when the fabrics are well-conditioned in actual use, but this could not be tested in this study. The nanoparticles were generated continuously during each filter test and the generator was paused and the airflow bypassed the filter for all velocity or filter changes.

10.6.1.1 Measurements An FMPS and an APS (Section 7.2.3) were operated simultaneously to record particle concentrations every second over the 5 nm to 20 μm diameter range. Normalized particle concentrations were measured upstream and downstream of the test filters; for each filter, two sets of concentration measurements over a period of 2 min (120 data points) per measurement were carried out for each upstream and downstream measurement and for each velocity.

None of the particles generated from the VENGES system were found to have diameters greater than 500 nm; therefore, the APS data were not used to calculate filtration efficiency. Aerosol particles were collected on TEM grids using sampling cassettes connected to the system as shown in Figure 7.16, using the method described in Sections 7.2.4 and 8.5. Transmission electron microscope (TEM)-copper grids (SPI 400 mesh with a Formvar/carbon film) were taped onto 37 mm diameter polycarbonate capillary-pore membrane filters (0.2 μm pore size). Air was drawn through the filters at 1 L/min using a calibrated personal sampling pump, and aerosol particles deposited on the grid via Brownian diffusion.

Particle concentration data including number, surface area, volume, mass and total concentrations recorded by the FMPS were exported to Excel spreadsheets for analysis. The collection efficiency of each filter was calculated for each individual channel (except the efficiency based on the total concentration) using Equation (9.3):

$$\eta_n = \frac{N_{in} - N_{out}}{N_{in}} \tag{9.3}$$

where, η_n = filter particle number collection efficiency, N_{in} = aerosol number concentration entering the filter (particles/cm^3), and N_{out} = aerosol number concentration leaving the filter (particles/cm^3).

Each concentration used in Equation 9.3 was the average of repeated 2-min samples (240 data points). Collection efficiency data profiles were statistically analyzed using SPSS. The Pearson correlation coefficient (γ) (Section 8.4.1) analyzed how collection efficiency profiles (η vs. particle diameter (d_p)) were associated with each other, for example, PF versus WP, or WP at 2.3 m/min velocity versus WP at 3.5 m/min velocity. Sampled particles on TEM grids were characterized following the methods described in Section 8.4.

Porosity Filter porosity was analyzed using a Quantachrome PoreMaster 33 mercury intrusion porosimeter, with data collected over a single intrusion/extrusion cycle. Filter samples were analyzed in glass cells with $0.5\,cm^3$ stem volumes, with sample size of squares approximately 1 cm by 1 cm. The contact angle and surface tension of the mercury were set to the default values of 140° (on intrusion and extrusion) and 480 erg/cm², respectively. For each sample, low-pressure (pneumatic) data were taken from the minimum starting pressure of approximately 1.4 kPa ($0.2\,lb/in^2$) up to 345 kPa ($50\,lb/in^2$), with all low-pressure data corrected versus a low-pressure blank run to minimize artifacts at the low starting pressures used. High-pressure data were taken from 140 kPa ($20\,lb/in^2$) up to 230,000 kPa ($33,000\,lb/in^2$), after which the two data sets were automatically merged using Quantachrome Poremaster for Windows 5.10; data from the intrusion cycle was used for calculation of average (volume median) pore size and pore size distribution in all cases. The overall range of analytical pressures used (~1.4 to 230,000 kPa) corresponded to an overall pore size range of approximately 1.1 mm to 6.5 nm.

Pressure Drop The pressure drop across each filter at both filtration velocities was measured by attaching a Dwyer inclined-vertical manometer across the upstream and downstream TEM sampling ports (Fig. 10.9). The pressure drop across the filter housing with no filter sample mounted was subtracted from the measured pressure drops to determine the actual pressure drop across each filter sample. The pressure drops were measured when the filters were new, and did not increase appreciably during the relatively short periods of particle collection used here.

10.6.1.2 Results

Generated Aerosols The typical aerosol particle size distributions generated by the VENGES system for testing filter media at 2.3 and 3.5 m/min face velocity are shown in Figure 10.10. Typically, when nanoparticles are dispersed as aerosols, they form agglomerates instead of the single particles in the primary size. This was consistently seen for SiO_2 particles produced by the VENGES system, and the mode of such agglomerates was about 100 nm for this study. The particle concentrations of up to 2×10^6 particles/cm³ at a particle diameter of 100 nm were sufficient to allow the calculation of collection efficiency greater than 99% for the high efficiency sampling filters at 2.3 m/min face velocity. The same sampling period, 120 s, was used for every test at downstream and upstream to provide a stable number of particles passing through the filters. The size range cut-off used to calculate size-specific collection efficiency was selected such that each size range contained at least 90% of the particles in the peak of the size distribution, which gave a range of 29–191 nm diameters for which collection efficiencies were calculated.

Filter Collection Efficiency   Collection efficiencies of all filters using metrics of number in each size channel of 29–191 nm, total number, total surface area, and total volume of 5–560 nm are presented in Table 10.1. The total collection efficiencies were calculated based on the total particle count by FMPS (5–560 nm). Rengasamy's study suggests that a particle number-based test method would be more applicable to reasonably ensure a certain level of filtration performance for a wide size range of particles including nanoparticles (Rengasamy et al., 2008). The collection efficiencies at two velocities based on number concentration over the

TABLE 10.1   Filter media collection efficiency (η) for 2.3 m/min face velocity and 3.5 m/min face velocity

	Collection efficiency (number %) Analysis by channel, 29–191 nm			Collection efficiency (%) Analysis of total count, 5–560 nm		
	η_{mean}	γ & *t*-test result	STD	Volume	Surface area	Number
2.3 m/min face velocity						
Quartz	98.3[a]	0.123, $p>0.05$	3.3	99.9	99.7	99.5
A/E	92.0[a]	0.165, $p>0.05$	13.8	99.7	99.7	99.5
WP-TMC	96.2[t]	0.284, $p<0.01$	1.3	97.0	96.9	96.4
PF-TMC	95.8[b,t]	0.842[**], $p<0.01$	2.4	93.7	94.4	95.6
PF-G	86.6[b,t]	−0.590, $p<0.01$	6.5	81.1	82.6	85.2
PF	64.4[c]	0.912[**], $p<0.01$	14.2	52.4	53.2	58.6
WP	49.8[c]	0.989[**], $p<0.01$	4.5	46.1	46.5	47.8
FP	36.2	−0.798[**], $p<0.01$	5.3	31.3	32.6	35.0
3.5 m/min face velocity						
Quartz	95.2[a]	0.123, $p>0.05$	4.9	94.7	94.3	93.1
A/E	94.4[a]	0.165, $p>0.05$	4.2	94.1	93.4	93.4
WP-TMC	74.1[t]	0.284, $p<0.01$	14.9	85.0	79.5	74.6
PF-TMC	51.0[t]	0.842[**], $p<0.01$	5.5	47.8	49.0	50.2
PF-G	65.1[t]	−0.590, $p<0.01$	1.3	65.7	65.8	64.2
PF	42.1	0.912[**], $p<0.01$	10.1	34.2	35.7	40.0
WP	27.8	0.989[**], $p<0.01$	2.9	25.7	26.0	25.8
FP	25.4	−0.798[**], $p<0.01$	3.6	27.7	27.4	25.3

η_{mean} % is the mean of number collection efficiencies from channels of 29–191 nm. $N=14$.

Superscript alphabet *a,b,c* are correlation results of "within group" comparison: Number collection efficiencies were statistically analyzed for the Pearson correlations among eight filters within the same velocity group. The efficiency numbers which share the same alphabet have high positive correlation ($\gamma>0.9$) to each other with significance at the 0.01 level.

Superscript *t* indicates that the paired *t*-test for "within group" comparison was tested, results showed that (a) at 2.3 m/min velocity, WP-TMC, and PF-TMC were not significantly different ($p=0.64$), and PF-TMC and PF-G are significantly different ($p<0.01$); (b) at 3.5 m/min velocity, WP-TMC to PF-TMC and PF-TMC to PF-G are significantly different ($p<0.01$), WP-TMC to PF-G are significantly different ($p<0.05$).

γ & *t*-test: Result of "between groups" comparison, Pearson correlation coefficient (γ) gives the correlation of collection efficiency and *p*-value of *t*-test gives the significant difference for each filter type between two groups of 2.3 and 3.5 m/min face velocity using number collection efficiency. $N=14$.

[**]Correlation is significant at the 0.01 level (2-tailed).

chosen aerosol diameters are shown in Figure 10.11. As shown in Table 10.1, the coated fabric filters, WP-TMC, PF-TMC, and PF-G, increased collection efficiencies by 20–30 percentage points versus the uncoated PF fabric filter and by 46 percentage points versus the uncoated WP filter, resulting in a collection efficiencies greater than 95% for the TMC fabric filters.

The collection efficiency based on total particle counts gave different results compared to average (mean) collection efficiency across the 29–191 nm channels due to the inclusion of the full FMPS size range of 5–560 nm. The efficiency of the sampling filters (quartz and A/E) increased for the full size range at 2.3 m/min face velocity, which indicates that the filters had higher collection efficiency for particle sizes above 191 nm and below 29 nm, but this increase was not seen at the 3.5 m/min face velocity. On the other hand, the efficiency of noncoated fabric filters tended to decrease for the full size range at both face velocities. However, the coated filters showed comparable collection efficiencies for the full size range and the selected size range and for both face velocities. The collection efficiency at 2.3 m/min was higher than at 3.5 m/min for all filters (except A/E); this was likely due to a longer residence time through the filtration region, giving more time for particles entrapment due to Brownian diffusion. At 2.3 m/min, for all fabrics except WP-TMC (which had a very high collection efficiency), the collection efficiency based on total number was higher than surface area, which was higher than volume; this indicates that the larger particles were collected with lower efficiency than the smaller ones, which is consistent with Figure 10.11a. This trend was reversed at the higher velocity for all fabrics except PF-TMC; this indicates that impaction was relatively more important at this velocity, since larger particles were collected preferentially. In general, the differences between number, surface area, and volume collection efficiencies were not great, indicating that collection efficiency did not vary greatly with particle size; this result is consistent with the relatively flat size-dependent collection efficiencies shown in Figure 10.11.

Statistical Analysis of Variability and Similarity of Filter Performance   Collection efficiency data profiles were analyzed using statistical tools as introduced above to support and validate the results; the Pearson correlation and *t*-test were used to statistically analyze data profile and the collection efficiency mean, respectively. The correlation showed high correlations for quartz and AE, PF-TMC and PFG, PF and WP. As shown in Figure 10.11, collection efficiency of these high-correlated filters at the 2.3 m/min face velocity mostly was inversely proportional to particle diameter less than 50 nm except the A/E filter. Therefore, the reduced efficiency of the A/E filter at particle diameters less than 50 nm at the 2.3 m/min velocity might be due to experimental error since the collection efficiency profile of the AE filter was statistically tested and had a statistically significant high correlation ($\gamma > 0.9$) to the quartz filter. The trend of increased efficiency at diameters less than 50 nm was not seen on most filters at the 3.5 m/min face velocity.

The most dominant filtration mechanism for nanoparticles is Brownian diffusion, and the smaller the particle size, the more Brownian diffusion will dominate the filtration mechanism (Hinds, 1999). The higher velocity would reduce the particle residence time in the filter, thus reducing the chance for collection by diffusion. The stability of filtration performance at 2.3 m/min velocity is high for aerosols in the range 29–191 nm. The WP-TMC filter tested at 3.5 m/min showed substantially reduced stability, while the nonwoven felted filters (PF-TMC, PF-G, PF) gave comparable stability.

Material Characterization and Filter Performance   The filter media were carefully characterized according to selected physical properties including fabric thickness, fabric weight, pressure drop, specific surface area (SSA), pore volume, and volume median pore size. Characterization results of all cleaned filter media and physical properties of selected filter media are shown in Table 10.2. Filter thickness and weight varied over a broad range; nonwoven fabrics were the thickest while the TMC and FP filter were the thinnest. WP filters had thicknesses in the same range as the quartz and A/E filters, but the weight was much reduced and the pressure drop was much lower than for the sampling filters. The SSA of the filters tended to increase when moving from clean to used filters for three filters whose SSA were measured, as shown in Table 10.2. The quartz filter consists solely of thin fibers and thus as expected has the highest SSA ($>4 \, m^2/g$), which is an order of magnitude higher than the SSA of WP-TMC and WP filters.

The particle collection time of all filters was the same so that comparable amounts of particles were collected by every filter. The SSA of the used WP-TMC filter increased about $0.25 \, m^2/g$ which was 74% higher than the clean WP-TMC filter, from 0.34 to $0.59 \, m^2/g$, and a similar increase in SSA was observed (about 19% from 0.27 to $0.32 \, m^2/g$) for the WP filter following use. On the other hand, the collected particles resulted in a $0.1–0.2 \, m^2/g$ SSA increase which would be barely detected on a high SSA filter media, thus, the collecting time for such filter needs to be extended to obtain a significant SSA increase. Therefore, the SSA change between clean and used quartz, which was about a 3% reduction, may not be significant due to the high overall surface area and could be due to a random variation between the measured areas. These increases are only observed when performing high-pressure mercury intrusion pressures; the low-pressure mercury intrusion data show no significant differences in the SSAs of the new and used filters, indicating that this difference comes from the presence of very fine pores or interstices between particles and consistent with nanoparticle entrapment in the filters.

In general, for fabrics the penetration can be expected to correlate directly with fiber diameter, pore volume, pore size, porosity and air permeability, and inversely with fabric thickness (Gao et al., 2011). The pore volume and pore size obtained in this study could not be correlated with filter efficiency. The fabric thickness affects the particle residence time, that is, the thicker the fabric, the longer the particle residence time. The FP woven filter was thinner than the other woven filters and gave the lowest collection efficiency under all conditions. In addition, uncoated woven fabrics such as the FP and WP tested in this study have large openings in their weave when new and are known to have very low collection efficiency until a dust layer is created on the surface of the filter (Strauss, 1975); this is consistent with the results found in this study. The uncoated PF had a much higher collection efficiency, which

TABLE 10.2 Filter media characterization results

Filter group	Material name	Structure	Thickness[a] (µm)	Weight[b] (mg)	SSA[c] (m²/g)	Pore volume[d] (cm³/g)	Volume median pore size[d] (µm)	ΔP[e](cm H_2O) (2.3 m/min)	(3.5 m/min)
Sampling filter	Quartz	Fine fiber	616±18	148	4.96, 4.82 (down 3%)	6	55	6.1	10.2
	A/E	Fine fiber	470±19	129	—	—	—	4.1	7.1
Coated fabric	WP-TMC	Woven fabric+Teflon fiber layer	679±26	566	0.34, 0.59 (up 74%)	1.2	30	1.5	3.6
	PF-TMC	Nonwoven fabric+Teflon fiber layer	2690±42	817	—	—	—	4.1	9.6
	PF-G	Nonwoven fabric+Goretex fiber layer	2030±23	883	—	—	—	3.8	6.1
Plain fabric	PF	Nonwoven fiber	1990±23	775	—	—	—	2.3	5.6
	WP	Woven fiber	883±26	677	0.27, 0.32 (up 19%)	1.8	75	0.8	1.5
	FP	Woven fiber	352±21	381	—	—	—	1.3	2.5

[a] Average of eight thickness measurements by micrometer ± one standard deviation.
[b] Weight was measured of filter samples with an area of 17.4 cm².
[c] Specific Surface Area (SSA) data are measurements taken on cleaned and used filter samples respectively in an area of 1 cm²; percent changes are in parentheses.
[d] Pore volume and volume median pore size data are for both clean and used filters taken in an area of 1 cm².
[e] Pressure Drop (ΔP) was corrected for ΔP across the empty filter housing.

is expected since felted fabrics rely on in-depth particle collection on individual fibers rather than the formation of a dust cake and thus should have high initial collection efficiencies when compared to woven fabrics.

This study has shown that fiber diameter plays an important role by demonstrating the strong influence of thin fiber diameter for collecting nanometer and submicrometer particles through Brownian diffusion. Since Brownian diffusion is the dominant filtration mechanism for nanoparticles, collection efficiency should not be affected by the particle density (Kim et al., 2007). Therefore, similar results are predicted when nanoparticles of different densities are collected in the filter media studied here.

In-line aerosol samples collected upstream and downstream of tested filters were analyzed by TEM. No silica ENPs were found on the downstream side of the high-efficiency sampling filters and the coated fabric filters. In contrast, many small ENP agglomerates were seen on the downstream side of the noncoated fabric filters with typical particles showing similar morphology to particles seen on the SEM images of used filters. Silica ENPs generated by VENGES were also collected for off-line characterization, the results indicating a specific surface area for the SiO_2 nanoparticles of $177\,m^2/g$, corresponding to an average SiO_2 primary particle size of $15.4\,nm$ with agglomeration clearly visible in TEM images.

The structures of the filter media and their collected nanoparticles were characterized by SEM. Typical structures for all studied filter media are shown in Figure 10.12. All used filters were tested at the 2.3 m/min filtration velocity; the collection duration for each filter was approximately 14 min, which was the total time of four measurements following the protocol. The aerosol concentration is shown in Figure 10.9. The quartz filter consists of a single type of thin fibers organized in a three-dimensional network as the representative structure of sampling filter (image A1 in Figure 10.12); ENPs were seen to be captured on the fibers (A2). The membrane coating layer consists of thin fiber structure attached to the top of fabric fibers. The Teflon membrane applied on the WP fiber (WP-TMC) showed a condensed and uniformed structure of thin Teflon fibers which were braided and connected with nodes (B1), while the Teflon membrane applied on the PF fiber (PF-TMC) showed more randomly braided Teflon fibers (D1). Both WP-TMC and PF-TMC fabric filters were fully covered by the Teflon fibers and both Teflon layers collected many NPs. The thinner these fibers, the more NPs were collected (B2, D2), which is consistent with filtration theory. The woven fabric filter (WP) shows bundles of woven fibers (C1) while the nonwoven (felted) filter (PF) shows fibers in a three-dimensional network structure (F1, F2). The Goretex membrane coating (E1) applied on the PF filter showed a much more open structure having many large nodes randomly mixed with Goretex fibers, and the felted fibers underneath the Goretex membrane layer were obvious in some areas. The used PF-G filter was able to collect many NPs on the Goretex fibers as well (E2).

Pressure Drop and Efficiency   Measured pressure drops and collection efficiency for the eight tested filters at the two filtration velocities are listed in Table 10.3. The results are in general agreement with filtration theory, which predicts that higher velocity results in higher pressure drop and lower collection efficiency.

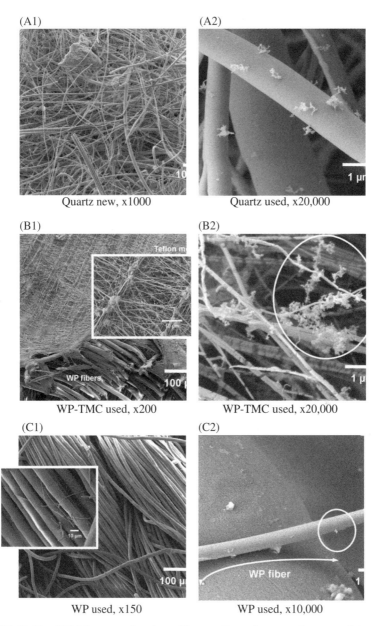

FIGURE 10.12 SEM images of various filter media and captured nanoparticles. (A1) Quartz—new filter, (A2) Quartz-used filter with NPs on fibers, (B1) WP-TMC—used filter showing Teflon membrane layer above the woven fibers at the cut edge, (B2) WP-TMC—used filter with NPs on Teflon fibers, (C1) WP—used filter showing woven fibers, (C2) WP used woven fibers with NPs on thin fiber,

FIGURE 10.12 (*Cont'd*) (D1) PF-TMC new filter showing Teflon membrane layer, (D2) PF-TMC—used filter with NPs on Teflon fibers, (E1) PF-G—new filter showing Goretex membrane layer above the nonwoven fibers, (E2) PF-G—used filter with NPs on Goretex fibers, (F1) PF-TMC—new filter showing the PF nonwoven fiber layer, and (F2) PF-TMC—used filter showing three-dimensional structure of felt fibers (Tsai et al., 2012). Used by permission, *Journal of Nanoparticle Research*.

TABLE 10.3   Data for pressure drop and collection efficiency at face velocities of 2.3 and 3.5 m/min for eight tested filters

	Pressure drop (cm H$_2$O)		Collection efficiency (%)	
	$V = 2.3$ m/min	$V = 3.5$ m/min	$V = 2.3$ m/min	$V = 3.5$ m/min
Quartz	6.1	10.2	98	95
A/E	4.1	7.1	92	94
WP-TMC	1.5	3.6	96	74
PF-TMC	4.1	9.7	96	51
PF-G	3.8	6.1	87	65
PF	2.3	5.6	64	42
WP	0.8	1.5	50	28
FP	1.3	2.5	36	25

Comparing the two air sampling filters to the six fabric filters, the air sampling filters have considerably higher pressure drop, which is not surprising since air sampling filters are optimized for collection efficiency with little concern for the resulting pressure drop. Fabric filters, on the other hand, must be designed to have both high collection efficiency and low pressure drop. Looking at the data for the uncoated fabrics, the two woven fabrics (WP, FP) have much lower pressure drops than the felted fabric (PF). This is expected, since the woven fabrics consist of a single layer of threads and the felt consists of random fibers in a thicker layer. In addition, it is expected that the coated version of a fabric would have a higher pressure drop than its uncoated counterpart; this was found for both the polyester felt and the woven polyester materials. Among the coated fabric filters, the polyester felt with a Teflon membrane coating (PF-TMC) had the highest pressure drop, with values approaching those measured for the two air sampling filters.

Summary   All filters had higher collection efficiency at the lower filtration velocity. The highest efficiency (>99.5%) was obtained using the quartz filter, followed by the glass fiber filter; this result confirms that air sampling filters are highly efficient in the nanometer size range. The performance of coated fabric filters was found to have a high variation between 50 and 96% efficiency; similar variation was seen for the uncoated fabric filters. A significant increase in collection efficiency was seen for coated fabric filters when compared to uncoated fabric filters. When tested at two velocities, both sampling filters maintained comparable performance according to the statistical analysis. In contrast, the performance of the fabric filters was significantly different under the different filtration velocities tested.

The sampling filters were designed to provide high collection efficiency under various conditions, and consistent results were seen in this study. On the other hand, environmental fabric filters gave much lower nanoparticle collection efficiency when operated under certain conditions. The 2.3 m/min filtration velocity is within the recommended range for both woven and nonwoven baghouse filters and was found to be an effective filtration velocity for all of the fabric filters tested in this study. According to the recommended velocity range for woven fabrics, the operating velocity would typically be below 2.3 m/min and theoretically the efficiency would be increased.

The filter media in each category showed unique performance characteristics. Coating layers such as Teflon and Goretex membrane, as expected, were found to have a major influence on fabric performance. Membrane layers constructed with more condensed and uniform thin fibers as seen in the Teflon membrane of WP and PF filters were able to successfully capture many silica nanoparticles and provided about 95% overall collection efficiency. In addition, the woven filters such as WP-TMC gave the best performance rating based on their low pressure drop operation, making them suitable for long-term use and lower energy consumption in the industry, especially given their light weight and low thickness.

In addition to fiber diameter, the structure of the membrane plays an important role. While the Goretex membrane was seen to collect many nanoparticles, it was not evenly coated on to the felted fibers, allowing nanoparticles to pass more easily through areas where fewer membrane fibers were present. For the uncoated fabric filters, the performance in capturing nanoparticles was low, as expected. Felted filters (PF) gave higher collection efficiencies than woven filters, with the three-dimensional structure of the felted fabric increasing the likelihood of capturing nanoparticles by diffusion, consistent with other studies published in the past (Bergmann, 1979; Culhane, 1974).

The collection efficiency of nanometer and submicrometer particles can be influenced by many filtration factors including filtration velocity, filter thickness, membrane coating, fiber diameter, pore size, pore volume, and porosity. In addition, the particle physical properties such as shape, size, and charge distribution, agglomeration, and surface area can influence the filter performance and require further research beyond this study. It is clearly seen from this work that an environmental fabric filter with a proper membrane coating such as the Teflon membranes described here can operate at very low pressure drop and give over 95% of collection efficiency for nanometer and submicrometer particles. A multi-filtration system using coated fabric filters as the primary filtration mechanism to be supplemented by a high efficiency filter such as a HEPA filter could be a practical and economic strategy for workplaces where significant amounts of nanometer and submicrometer particles are emitted from production processes.

10.6.2 The Challenge and Brainstorming

Paul has been working with the CNT furnace placed in a fume hood for few years in a research lab. Recently, the EHS officer examined the hood and the activities performed in the hood.

10.6.3 Study Questions

1. What situations are suitable for using a HEPA filter? Can such a filter be used here?

2. What control methods can be applied to a CNT furnace?
3. Is a fume hood likely to provide effective control for this process? Should some other type of custom-designed ventilated enclosure be used?

REFERENCES

Baalousha M, Lead JR, von der Kammer F, Hofmann T. Natural colloids and nanoparticles in aquatic and terrestrial environments. In: Lead JR, Smith E, editors. *Environmental and Human Health Impacts of Nanotechnology*. Chichester: John Wiley & Sons, Ltd.; 2009.

Benn TM, Westerhoff P. Nanoparticle silver released into water from commercially available sock fabrics. Environ Sci Technol 2008;42:4133–4139.

Bergmann L. Needled felts for fabric filters. Environ Sci Technol 1979;13:1472–1475.

Calvert S. Venturi and other atomizing scrubbers efficiency and pressure drop. AIChE J 1970;16:392–396.

Chen BT, Afshari A, Stone S, Jackson M, Schwegler-Berry D, Frazer DG, Castranova V, Thomas TA. Nanoparticles-containing spray can aerosol: characterization, exposure assessment, and generator design. Inhal Toxicol 2010;22:1072–1082.

Culhane FR. Fabric filters abate air emissions. Environ Sci Technol 1974;8:127–130.

Danielson JA. *Air Pollution Engineering Manual AP-40*. Washington: U.S. Environmental Protection Agency; 1973.

Demokritou P, Buchel R, Molina RM, Deloid G, Brain J, Pratsinis SE. Development and characterization of a Versatile Engineered Nanomaterial Generation System (VENGES) suitable for toxicological studies. Inhal Toxicol 2010;22:107–116.

Ellenbecker M, Tsai C, Isaacs J. *Best Practices for Working Safely with Nanoparticles in University Research Laboratories*. Lowell, MA: University of Massachusetts Lowell; 2008.

Gao P, Jaques PA, Hsiao TC, Shepherd A, Eimer BC, Yang M, Miller A, Gupta B, Shaffer R. Evaluation of nano- and submicron particle penetration through ten nonwoven fabrics using a wind-driven approach. J Occup Environ Hyg 2011;8:13–22.

Geranio L, Heuberger M, Nowack B. The behavior of silver nanotextiles during washing. Environ Sci Technol 2009;43:8113–8118.

Goyal D, Zhang XJ, Rooney-Varga JN. Impacts of single-walled carbon nanotubes on microbial community structure in activated sludge. Lett Appl Microbiol 2010;51:428–435.

Hallock MF, Greenley P, DiBerardinis L, Kallin D. Potential risks of nanomaterials and how to safely handle materials of uncertain toxicity. J Chem Health Safety 2009;January–February:16–23.

Hinds WC. *Aerosol Technology—Properties, Behavior, and Measurement of Airborne Particles*. New York: Wiley-Interscience; 1999.

Kim SC, Harrington MS, Pui DYH. Experimental study of nanoparticles penetration through commercial filter media. J Nanopart Res 2007;9:117–125.

Kulthong K, Srisung S, Boonpavanitchakul K, Kangwansupamonkon W, Maniratanachote R. Determination of silver nanoparticle release from antibacterial fabrics into artificial sweat. Part Fibre Toxicol 2010;7:8.

Limbach LK, Bereiter R, Muller E, Krebs R, Galli R, Stark WJ. Removal of oxide nanoparticles in a model wastewater treatment plant: influence of agglomeration and surfactants on clearing efficiency. Environ Sci Technol 2008;42:5828–5833.

Lin GU, Chen TM, Tsai CJ. A modified Deutsch-Anderson equation for predicting the nanoparticle collection efficiency of electrostatic precipitators. Aerosol Air Qual Res 2012;12:697–706.

Luongo AL, Zhang X. Toxicity of carbon nanotubes to the activated sludge process. J Hazard Mater 2010;178:356–362.

Mädler L, Friedlander SK. Transport of nanoparticles in gases: overview and recent advances. Aerosol Air Qual Res 2007;7:304–342.

Mohr M, Matter D, Burtscher H. Efficient multiple charging of diesel particles by photoemission. Aerosol Sci Technol 2007;24:14–20.

Mu H, Chen Y. Long-term effect of ZnO nanoparticles on waste activated sludge anaerobic digestion. Water Res 2011;45:5612–5620.

Nazarenko Y, Han TW, Lioy PJ, Mainelis G. Potential for exposure to engineered nanoparticles from nanotechnology-based consumer spray products. J Expo Sci Environ Epidemiol 2011;21:515–528.

Nazarenko Y, Zhen H, Han T, Lioy PJ, Mainelis G. Potential for inhalation exposure to engineered nanoparticles from nanotechnology-based cosmetic powders. Environ Health Perspect 2012;120:885–892.

Pui DYH, Fruin S, McMurry PH. Unipolar diffusion charging of ultrafine aerosols. Aerosol Sci Technol 1988;8:173–187.

Reijinders L. Cleaner nanotechnology and hazard reduction of manufactured nanoparticles. J Clean Prod 2006;14:124–133.

Rengasamy S, King WP, Eimer BC, Shaffer RE. Filtration performance of NIOSH-approved N95 and P100 filtering facepiece respirators against 4 to 30 nanometer-size nanoparticles. J Occup Environ Hyg 2008;5:556–564.

Strauss W. *Industrial Gas Cleaning*. New York: Pergamon Press; 1975.

Tsai CJ, Lin CH, Wang UM, Hunag CH, Li SN, Wu ZX, Wang FC. An efficient venturi scrubber system to remove submicron particles in exhaust gas. J Air Waste Manage Assoc 2005;55:319–325.

Tsai SJ, Echevarría-Vega M, Sotiriou G, Santeufemio C, Huang C, Schmidt D, Demokritou P, Ellenbecker M. Evaluation of environmental filtration control of engineered nanoparticles using the Harvard Versatile Engineered Nanomaterial Generation System (VENGES). J Nanopart Res 2012;14:812.

University of Pennsylvania. Nanoparticle handling fact sheet. Philadelphia (PA): University of Pennsylvania; 2008.

Virginia Commonwealth University [VCU]. *Nanotechnology and Nanoparticles: Safe Working Practices Information Page*. Richmond, VA: Virginia Commonwealth University; 2007.

Wiesner MR, Lowry GV, Alvarez P, Dionysiou D, Biswas P. Assessing the risks of manufactured nanomaterials. Environ Sci Technol 2006;40:4336–4345.

Yin Y, Zhang X, Graham J, Luongo L. Examination of purified single-walled carbon nanotubes on activated sludge process using batch reactors. J Environ Sci Health A 2009;44:661–665.

11

THE REGULATORY ENVIRONMENT FOR ENGINEERED NANOMATERIALS

Since about 1970, the number of occupational and environmental health laws and regulations has proliferated, both in the United States and throughout the world. A list of the acronyms of important US federal regulations and agencies would include EPA, OSHA (the agency and the Act), NIOSH, CAA, CWA, FIFRA, FDA, CPSC, MSHA, TSCA, CERCLA, RCRA—the alphabet soup goes on and on. A similar list can be assembled for the EU and other industrialized countries. This naturally leads us to the question—how do all of these regulations apply to the emerging field of nanotechnology? The short answer is—not very well. Most environmental laws and regulations were developed in response to hazards that were known at the time they were written, and apply poorly or not at all to new hazards.

Most of the major US environmental and occupational health regulations apply only to a specific list of chemicals and those lists are updated very slowly. For example, the Occupational Safety and Health Act (discussed below) includes a list of chemicals that have permissible exposure limits (PELs) (OSHA, 2006). The Act includes provisions for changing the list, but in more than 40 years since the Act's passage, the list has been modified less than once per year. Based on this very slow response to new hazards, it is unlikely that OSHA will issue PELs for engineered nanomaterials (ENMs) for many years to come. However, the Occupational Safety and Health Act articulates a general duty to provide a safe workplace (OSHA, 2007); therefore the absence of specific rules does not mean that the risks of using nanomaterials can be ignored.

Exposure Assessment and Safety Considerations for Working with Engineered Nanoparticles, First Edition. Michael J. Ellenbecker and Candace Su-Jung Tsai.
© 2015 John Wiley & Sons, Inc. Published 2015 by John Wiley & Sons, Inc.

In addition, several laws, regulations, and agencies in the United States and the EU do have some direct relevance to nanotechnology. This chapter will review the major existing laws that impact this field, while Chapter 12 will include recommendations for additional future regulations.

11.1   OCCUPATIONAL HEALTH REGULATIONS

Regulations regarding worker exposure to toxic materials have been promulgated by the federal governments in all industrialized countries. With regard to ENMs, it is fair to say that as of this writing, the governments of most industrialized countries have focused on research and policy recommendations, rather than specific regulations.

11.1.1   Occupational Health Regulations in the European Union

On June 12, 1989, the European Union (EU) issued Framework Directive 89/391/EEC, concerning general rules for health and safety at work (Commission of the European Communities, 1989). This directive applies to the manufacturing and use of chemicals throughout the production process, regardless of the number of workers or the quantity of the chemical involved. According to the document titled "Regulatory Aspects of Nanomaterials" (Commission of the European Communities, 2008a):

> This Directive fully applies to nanomaterials. Employers, therefore, must carry out a risk assessment and, where a risk is identified, take measures to eliminate this risk.
>
> The planning and introduction of new technologies must be subject to consultation with the workers or their representatives, as regards the working conditions and the working environment in accordance with Articles 11 and 12 of the Framework Directive 89/391/EEC.
>
> The Directive foresees the possibility to adopt individual directives laying down more specific provisions with respect to particular aspects of safety and health. Relevant directives thus adopted relate to risks related to exposure to carcinogens or mutagens at work, risks related to chemical agents at work, the use of work equipment by workers at work, the use of personal protective equipment at the workplace and safety and health protection of workers potentially at risk from explosive atmospheres.
>
> As these Directives introduce minimum requirements, national authorities have the possibility to introduce more stringent rules.

In 1998, the EU issued Council Directive 98/24/EC (Commission of the European Communities, 1998), a directive within the broader Directive 89/391/EEC, addressed specifically to the protection of the health and safety of workers from chemical hazards. Its provisions, which also apply to ENMs, include very specific and comprehensive requirements for risk assessment, risk prevention, and health surveillance.

A few specific nanomaterial occupational exposure limits (OELs) have been issued within the EU. With regard to carbon nanotubes (CNTs), the British Standards Institute

(BSI) in 2007 recommended a "benchmark" OEL of 0.01 fibers/cm^3, as measured by transmission or scanning electron microscopy; this is equivalent to the most rigorous exposure limit in Britain for asbestos, that is, the clearance limit for asbestos removal activities (the same limit as used by the US EPA for this activity) (BSI, 2007). This recommended benchmark takes a different approach from the NIOSH draft method discussed in the following, in that it measures actual airborne CNT concentrations and represents a much lower allowable CNT exposure than that of the draft NIOSH method.

The German company Bayer Schering Pharmaceuticals has studied the toxicity of their multiwalled CNTs called Baytubes. Their conclusion is that they are not likely to lead to mesothelioma or other chronic conditions because their flexibility leads to the formation of relatively large assemblages, or "bird's nests" of the tubes; based on measured acute toxicity in rats, they have set a company OEL of 50 μg/m^3 for Baytubes (Pauluhn, 2010).

Given the worldwide shortage of ENP OELs, and the lengthy time needed for their adoption, van Broekhuizen and Dorbeck-Jung (2013) describe a possible voluntary approach that can serve in the interim. The German Institute for Occupational Safety and Health (IFA) has developed what are called benchmark levels of evaluating ENP exposures. The benchmarks use what IFA considers to be likely predictors of ENP toxicity, that is, size, shape, density, and biopersistence, to distinguish four groups, each with a "nano reference value (NRV)." Group 1 consists of "rigid, biopersistent nanofibers for which effects similar to those of asbestos are not excluded" (e.g., CNTs) with a NRV of 0.01 fibers/cm^3 (the BSI standard). Group 2 includes "biopersistent granular nanomaterial in the range of 1 and 100 nm" with a density greater than 6000 kg/m^3 (e.g., gold, silver) with a NRV of 20,000 particles/cm^3. Group 3 is similar to Group 2, but for materials with a density less than 6000 kg/m^3 (e.g., TiO_2, ZnO) with a NRV of 40,000 particles/cm^3. The Group 2 and 3 NRVs are meant to correspond to a mass concentration of 0.1 mg/m^3. Group 4, "non-biopersistent granular nanomaterial in the range of 1 and 100 nm" (e.g., sodium chloride) is considered to have toxicity similar to the material at larger sizes, so the applicable OEL for that material is used.

van Broekhuizen and Dorbeck-Jung (2013) emphasize the voluntary nature of this approach:

NRVs are intended to be precautionary warning levels: when they are exceeded, exposure control measures should be taken. As such, they support compliance with the legal duty to control the health risks of MNMs. Use of NRVs requires measurement of the particle concentration and diameter and requires limited information about the identity of the processed (and measured) MNMs. For identification, information is required about the shape of the MNMs (fiber or sphere-like shape), its biopersistency, and information on the density of the nanomaterial.

Concurrently, NRVs are not legally binding. By regarding NRVs as part of the current state of science the Dutch Minister of Social Affairs and Employment has recommended the use of NRVs as provisional limit values that should be accompanied by additional measures to minimize exposure. The Minister's recommendation can be regarded as a "soft" regulation. Although not legally binding, this regulatory measure involves certain commitments to either employ the NRVs or search for alternatives.

The authors describe a pilot study conducted in 2010 "to investigate whether NRVs are accepted in practice and how their usefulness is perceived." They conclude that "most companies working with nanomaterials accept NRVs as a tool to minimize possible adverse health effects among employees. Companies tend to be pro-active and acquiescent toward using the NRVs for risk assessment and management."

11.1.2   US Occupational Health Regulations

11.1.2.1   Occupational Exposure Limits   In the United States, the Occupational Safety and Health Agency (OSHA), established by the Occupational Safety and Health Act of 1970, is responsible for protecting the health of workers. With regard to toxic materials, OSHA's principal tool for filling this role is the promulgation of PELs, defined as the maximum allowable airborne exposure for that particular material (OSHA, 2006). Most PELs are 8-h time-weighted averages, although materials that can cause short-term acute effects may have a ceiling value that cannot be exceeded even for a short time.

As of this writing, no PELs specific to ENMs have been promulgated; thus, OSHA can only regulate exposures under Section 5(a)(1) of the Act (OSHA, 1970), commonly known as the general duty clause, which requires employers to "furnish to each of his employees employment and a place of employment which are free from recognized hazards that are causing or are likely to cause death or serious physical harm to his employees." Statements by OSHA personnel at recent meetings have indicated that OSHA has yet to issue a citation under the general duty clause for exposure to an ENM.

The National Institute for Occupational Safety and Health (NIOSH) was also established by the Act and has responsibility for performing research on occupational health hazards. If NIOSH determines that there is sufficient evidence for OSHA to establish a new or revised PEL, they issue a recommended exposure limit (REL), publish it, and forward it to OSHA. To date, two RELs, for titanium dioxide (TiO_2) and CNTs, have been published.

NIOSH proposes RELs in documents called Current Intelligence Bulletins. Current Intelligence Bulletin 63 addresses TiO_2 (NIOSH, 2011). In this document, NIOSH defines *fine* TiO_2 particles as those meeting the respirable particle sampling definition (see Section 4.2.2) and *ultrafine* TiO_2 particles as those with diameters less than 100 nm; this is also called nanoparticle TiO_2. After reviewing the extensive documentation on TiO_2 toxicity, NIOSH proposed separate RELs for the fine and ultrafine forms (NIOSH, 2011):

NIOSH recommends airborne exposure limits of 2.4 mg/m³ for fine TiO_2 and 0.3 mg/m³ for ultrafine (including engineered nanoscale) TiO_2, as time-weighted average (TWA) concentrations for up to 10 hr/day during a 40-hour work week. These recommendations represent levels that over a working lifetime are estimated to reduce risks of lung cancer to below 1 in 1,000. The recommendations are based on using chronic inhalation studies in rats to predict lung tumor risks in humans.

Note that the REL for ultrafine TiO_2 is considerably lower than that for fine TiO_2; they are both much lower than the current OSHA PEL of $15\,mg/m^3$, which is a total dust standard.

In 2010, a second REL, for CNTs and carbon nanofibers (CNFs), was published as a draft Current Intelligence Bulletin (NIOSH, 2010), and comments from interested parties were solicited. The CIB reviewed in detail the toxicology literature concerning the acute health effects associated with CNTs and CNFs (see Section 5.4), and proposed an REL of $7\,\mu g/m^3$ of airborne elemental carbon as an 8-h TWA respirable mass concentration. Exposures are to be measured and compared to the REL using NIOSH Method 5040, titled "Diesel Particulate Matter (as Elemental Carbon)" (NIOSH, 2003).

As indicated by the title, this particular NIOSH analytical method was developed to measure elemental carbon contained in diesel particulate, but the draft CIB makes the case that elemental carbon can be used as a surrogate for actually measuring CNTs or CNFs, since they are largely composed of elemental carbon (EC). The $7\,\mu g/m^3$ exposure level was selected based on a combination of two factors. First, this is the published upper estimate of the limit of quantitation (LOQ) for Method 5040 when sampling for 8h using a personal sampling pump, where LOQ is defined as the lowest quantity of a substance that can be distinguished from the absence of that substance by the analytical method. Second, the NIOSH risk assessment, described in detail in the draft CIB, concluded that exposure at or below this level will "…minimize the risk for developing adverse lung effects" (NIOSH, 2010).

NIOSH received numerous comments in response to the draft Current Intelligence Bulletin and published the final Current Intelligence Bulletin 65 in April 2013 (NIOSH, 2013). They reviewed all of the new toxicology literature that had been published since the release of the draft Current Intelligence Bulletin. Most importantly, they reassessed the stated limit of detection (LOD) and LOQ of Method 5040. The LOD is the statistically significant minimum quantity (in this case, EC mass concentration) that can be detected when compared to a blank filter, while the LOQ is the statistically significant minimum concentration that can be quantified. A common layperson's description of the difference is that, in a crowded room, the LOD is the minimum voice amplitude that can be *heard* above the background noise, while the LOQ is the minimum amplitude that can be *understood* above the background noise.

In any case, the Current Intelligence Bulletin goes through a lengthy description of the experimental evidence justifying the lowering of the LOD to $0.1\,\mu g/m^3$ and the LOQ to $1\,\mu g/m^3$. For typical concentrations, NIOSH estimates that "under optimum conditions an LOQ of $1\,\mu g/m^3$ can be obtained for an 8-hr respirable sample collected on a 25-mm filter at a flow rate of 4 liters per minute (lpm)."

For a 45-year lifetime exposure at the REL of $1\,\mu g/m^3$ (8-h time-weighted average), NIOSH estimates that the "maximum likelihood estimate" of "minimal lung effects" as 2.4–33% and the maximum likelihood estimate of "slight or mild lung effects" as 0.23–10%. They then state that "NIOSH does not consider a 10% estimated excess risk over a working lifetime to be acceptable for these early-stage lung effects, and the REL is set at the optimal limit of quantification (LOQ) of the analytical method carbon" (NIOSH, 2003).

Although lowering the REL from 7 to 1 µg/m³ resulted in a recommended standard that is more protective of workers, concerns remain. The first major concern is that the REL, as recognized by NIOSH in the previous paragraph, may not actually be protective of worker health. In addition, the risk assessment and the REL did not consider mesothelioma as an end point, even though there is considerable animal data indicating that this is a distinct possibility (see Section 5.4).

A second major concern is the fact that NIOSH is proposing a mass-based exposure limit for CNTs and CNFs. The primary reason for doing so is that standard industrial hygiene sampling equipment (i.e., personal sampling pump and filter cassette) and an existing analytical method can be used to evaluate exposures and compare them to the exposure limit. However, as discussed in some detail in Section 3.3, very small mass concentrations of nanoparticles can represent a very high number concentration. This can be demonstrated by a simple calculation. Assume that an occupational hygienist is evaluating an exposure to CNTs using NIOSH Method 5040 and finds an elemental carbon concentration of 1 µg/m³, just equal to the NIOSH REL. Assume further that the measured aerosol consisted of CNTs with a diameter (d) of 10 nm, a length (L) of 2000 nm, and a density (ρ) of 1.4 g/cm³. We can first calculate the volume of one such CNT:

$$V = \frac{\pi d^2 L}{4} = \frac{\pi (10^{-8}\,\mathrm{m})^2 (2 \times 10^{-6}\,\mathrm{m})}{4} = 1.6 \times 10^{-22}\,\mathrm{m}^3$$

The mass of this CNT is:

$$m = \rho V = (1400\,\mathrm{kg/m^3})(1.6 \times 10^{-22}\,\mathrm{m}^3)$$
$$= 2.2 \times 10^{-19}\,\mathrm{kg} \times 10^{9}\,\mathrm{µg/kg} = 2.2 \times 10^{-10}\,\mathrm{µg}$$

We can now convert from the mass concentration of the REL to the equivalent number concentration:

$$n = \frac{c}{m} = \frac{1\,\mathrm{µg/m^3}}{2.2 \times 10^{-10}\,\mathrm{µg/CNT}} = 4.6 \times 10^{9}\,\mathrm{CNT/m^3} \times 10^{-6}\,\mathrm{m^3/cm^3} = 4600\,\mathrm{CNT/cm^3}$$

This is an extremely high number concentration, especially when compared for examples to asbestos, which we know causes mesothelioma and has a PEL of 0.1 fibers/cm³. The simple assumptions in this example might of course be entirely unrealistic. For example, most reports of actual monitoring of CNTs in the workplace (Han et al., 2008; Myojo et al., 2009; Tsai et al., 2009) have not found individual CNTs, but a mixture of amorphous carbon "soot" nanoparticles and CNT bundles. But even if 90% of the measured mass concentration was soot, and the remaining 10% consisted of bundles of CNTs with an average of 100 CNTs/bundle, there still would be 4–5 such bundles/cm³ at the REL.

On the other hand, a CNT manufacturing process may release a considerable number of carbon nanoparticles, or soot, but no actual CNTs. In this case, a sample collected using NIOSH Method 5040 may show a high airborne carbon mass concentration, but this will not represent a possible CNT exposure. Of course, carbon nanoparticles may also be hazardous (Oberdörster et al., 2007), but the method is meant to measure CNT exposures. The inability of Method 5040 to distinguish between CNTs and other particles containing elemental carbon is another disadvantage of this approach.

Another variable is the size of the CNTs being sampled. A calculation similar to that above was performed in Section 7.2, this time using much longer but thinner SWCNTs, and the resulting concentration, correcting for the difference in mass concentration in the two examples, is $500 CNT/cm^3$.

The NIOSH CIB does recognize and acknowledge these difficulties with the proposed REL, stating:

> The recommended exposure limit is in units of mass/unit volume of air, which is how the exposures in the animal studies were quantified and it is the exposure metric that generally is used in the practice of industrial hygiene. In the future, as more data are obtained, a recommended exposure limit might be based on a different exposure metric better correlated with toxicological effects, such as CNT/CNF number concentration.

The Current Intelligence Bulletin also contains a great deal of information on engineering controls, respiratory protection, and good work practices to minimize worker exposure (similar to what is discussed elsewhere in this book) while further work is done to refine the CNT REL.

There appears to be a clear difference between the EU and US approaches to setting ENM OELs. The Europeans, as discussed earlier, are moving toward the setting of number-based standards for the more hazardous ENMs such as CNTs, while the United States is moving toward a mass-based OEL. This has implications for future research, since it will not be possible to directly compare and collect the data from studies performed in the two areas. Such data combination may be useful, for example, when assembling a large collection of workers for an epidemiology study. This issue is discussed further in Section 12.2.

The American Conference of Governmental Industrial Hygienists (ACGIH) was formed in the late 1940s, with the express purpose of promulgating occupational exposure limits (OELs). ACGIH published its first list of OELs, called threshold limit values (TLVs®), in 1946; from that date until OSHA and NIOSH were created in 1970, the TLVs were the primary OEL values used throughout the world, and they still are widely referenced today. Since the TLVs are set by the ACGIH TLV committee, which does not have to clear all of the regulatory hurdles faced by OSHA and NIOSH, it is generally thought that TLVs can be established faster than PELs or RELs; thus, there is some hope in the nanotechnology EHS community that ACGIH may be a source for ENM OELs in the near future. However, at a recent meeting, the head of the ACGIH TLV committee stated that they are not now studying ENMs and that they "may" do so in the future, so it may be quite some time before we see any such TLVs.

11.1.2.2 Hazard Communication In order to comply with OSHA's Hazard Communication Standard, employers must identify all hazards associated with the material and communicate them to employees, along with information on how they may protect themselves from harm. In guidance developed to assist in compliance with the standard (OSHA), OSHA explains that the hazard determination required by the act is not a risk determination:

> Hazard determination is the process of evaluating available scientific evidence in order to determine if a chemical is hazardous pursuant to the HCS (Hazard Communication Standard). This evaluation identifies both physical hazards (e.g., flammability or reactivity) and health hazards (e.g., carcinogenicity or sensitization). The hazard determination provides the basis for the hazard information that is provided in MSDSs [material safety data sheets], labels, and employee training.
>
> Hazard determination does not involve an estimation of risk. The difference between the terms hazard and risk is often poorly understood. Hazard refers to an inherent property of a substance that is capable of causing an adverse effect. Risk, on the other hand, refers to the probability that an adverse effect will occur with specific exposure conditions. Thus, a substance will present the same hazard in all situations due to its innate chemical or physical properties and its actions on cells and tissues. However, considerable differences may exist in the risk posed by a substance, depending on how the substance is contained or handled, personal protective measures used, and other conditions that result in or limit exposure. This document addresses only the hazard determination process, and will not discuss risk assessment, which is not performed under the OSHA HCS.

Using the principles outlined above, MSDSs should describe the increased hazard of ultrafine titanium oxide nanoparticles as compared to fine, and the identified hazards of CNTs should be identified. In addition, good work practices as recommended by NIOSH and others should be communicated to workers. Manufacturers and importers responsible for providing MSDSs to purchasers should not wait for PELs to be developed to include this information, nor should employers be satisfied with MSDSs received from manufacturers that have not identified these recognized hazards. Unfortunately, as discussed in Section 9.3, many MSDSs for ENPs have failed to adequately describe their hazards.

11.1.3 Summary: Occupational Exposure Regulations

At this time, we are faced with a highly unsatisfactory combination of factors. Research groups and manufacturers are using increasing quantities of ENMs, potentially exposing more and more workers to higher levels of these materials. At the same time, toxicology studies on these materials have advanced slowly, so that only a handful of OELs have been issued by government and private groups. Consequently, individuals responsible for protecting the health of nanotechnology workers have little information as to what levels of exposure should cause them concern.

It gets worse. There is no agreement in the scientific community on the *metrics* to be used in future OELs. Despite the difficulties in collecting mass samples for ENMs, which are described in some detail in Section 7.2.1, organizations such as NIOSH and Bayer, among the few existing OELS described earlier, are employing mass sampling. A major reason for this is that the equipment now available to measure particle count and/or size distribution in the nanometer size range is expensive, bulky, and difficult to use. Such instruments are not appropriate for a field occupational hygienist to use in measuring against an OEL. This problem is discussed further in Section 12.3, where we conclude the book by discussing future regulatory needs. In the meantime, voluntary approaches such as the use of NRVs may be a promising means to help fill the regulatory gap.

11.2 ENVIRONMENTAL REGULATIONS

At this time, the state of environmental regulations relative to ENMs is very similar to that of occupational regulations—regulatory agencies are *studying* the problem and issuing interim *recommendations*, with very few actual regulations having been promulgated. However, general responsibilities may apply in a manner that is often less than clear, but which can result in liabilities if insufficient attention is paid. This section will first review the status of environmental regulations in the United States, followed by a brief discussion of Europe and the rest of the world.

11.2.1 US Environmental Regulations

US environmental regulations are the purview of the Environmental Protection Agency (EPA), which has been moving slowly toward the regulation of ENMs. EPA initially focused on voluntary efforts, by establishing the Nanoscale Materials Stewardship Program (NMSP) in January 2008. According to EPA (2009):

The NMSP has two components, a Basic Program and an In-Depth Program:

- Under the Basic Program, EPA asked manufacturers, processors, importers, and users of chemical nanoscale materials to submit existing information on their nanoscale materials, i.e., physical and chemical properties, hazard, exposure, use, and risk management practices or plans, during the first six months of the program. The Agency also invited researchers who develop or study engineered nanoscale materials to participate.
- Under the In-Depth Program, EPA invited participants to sponsor the development of test data for representative nanoscale materials and to work with the Agency to devise a data development plan. A particular sponsor may choose to implement one or more aspects of the plan; or a consortium of sponsors and other stake-holders may work together to implement aspects of the plan. At its completion, EPA and sponsors will review the information gathered; consider input from stakeholders; conduct final assessments; and consider further action.

EPA originally set July 28, 2008, as the date by which participants in the Basic Program should report information on the ENMs they manufacture, import, process, or use. By that date, EPA received submissions from only 16 companies and trade associations; those submissions reported information of 91 different nanoscale materials based on 47 different chemicals. EPA continued to accept information from participants into 2010 and has published an updated list on their NMSP web site.[1] At the conclusion of the program, they had received information from 31 organizations covering "more than 132 nanoscale materials." In its January 2009 interim report on the program (EPA, 2009), the Agency concluded:

> Thus far, the program has considerably increased the Agency's understanding of the types of nanoscale materials in commerce. Most submissions included information on physical and chemical properties, commercial use (realized or projected), basic manufacturing and processes as well as risk management practices. However, very few submissions provided either toxicity or fate studies. Because many submitters claimed some information as confidential business information, the Agency is limited in the details of what it can report for any particular submission.

11.2.1.1 Toxics Substance Control Act Beyond information gathering, as reflected in the NMSP, EPA has focused most of its actual regulatory actions on the Toxics Substance Control Act (TSCA). When TSCA was passed and became effective in 1977, it established a Chemical Substance Inventory of chemicals considered to be "existing" as of that date. Any substance not already included on the Inventory is considered to be a "new" chemical and anyone wishing to manufacture or import a "new" chemical must submit a Premanufacture Notice (PMN) to EPA at least 90 days in advance. A PMN must include information on production volume, anticipated uses, and potential for release and exposure, to the extent that this information is known at the premanufacture stage. Importantly, no information is required regarding the material's toxicity or environmental fate and transport. Upon review and approval by EPA, the "new" chemical is added to the Inventory and becomes an "existing" chemical substance.

As discussed throughout this book, simply reducing the particle size of a material to the nanometer range often gives it new properties, including perhaps new toxicity; almost implicitly, we have been considering such nanoparticles to be "new" materials. Given that, a question of great concern has arisen in the ENM community regarding which ENMs, if any, must file PMN notices under TSCA as a "new" material. EPA has attempted to give guidance on this question (EPA, 2008a):

> Section 3(2)(A) of TSCA defines the term "chemical substance" to mean "any organic or inorganic substance of a particular molecular identity...." Thus, in determining whether a chemical substance is a new chemical for purposes of TSCA Section 5, or instead is an existing chemical, EPA determines whether the chemical substance has the same molecular identity as a substance already on the inventory. A chemical substance with a molecular identity that is not identical to any chemical substance on the TSCA Inventory is considered to be a new chemical substance (i.e. not on the inventory); a chemical substance that has the same molecular identity as a substance listed on the Inventory is considered to be an existing chemical substance.

[1] http://www.epa.gov/oppt/nano/stewardship.htm

EPA views molecular identity as being based on such structural and compositional features as the types and number of atoms in the molecule, the types and number of chemical bonds, the connectivity of the atoms in the molecule, and the spatial arrangement of the atoms within the molecule. EPA considers chemical substances that differ in any of these structural and compositional features to have different molecular identities. For example, EPA considers chemical substances to have different molecular identities for the purposes of TSCA when they:

- have different molecular formulas, i.e., they have the same types of atoms but a different number of atoms, e.g., ethane (C_2H_6) and propane (C_3H_8), or they have the same number of atoms but different types of atoms, e.g., bromomethane (CH_3Br) and chloromethane (CH_3Cl), or they differ in both the types and numbers of atoms.
- have the same molecular formulas but have different atom connectivities, i.e., they have the same types and number of atoms but are structural isomers (e.g., n-butane and isobutane) or positional isomers (e.g., 1-butanol and 2-butanol).
- have the same molecular formulas and atom connectivities but have different spatial arrangements of atoms, e.g., they have the same types, number, and connectivity of atoms but are isomeric (e.g., (Z)-2-butene and (E)-2-butene).
- have the same types of atoms but have different crystal lattices, i.e., they have different spatial arrangements of the atoms comprising the crystals, e.g., anatase (atoms arrayed tetragonally) and brookite (atoms arrayed orthorhombically) forms of titanium dioxide.
- are different allotropes of the same element, e.g., graphite (carbon atoms arranged in hexagonal sheets with each atom bonded to three other atoms in the plane of a given sheet) and diamond (carbon atoms arranged in a tetrahedral lattice with each atom bonded to four other atoms).
- have different isotopes of the same elements.

Given the above definitions, it should be possible to categorize all ENMs as either "new" or "existing" chemicals. For example, graphite, a form of carbon, has the same molecular formula, atom connectivity, and so on, in the nanometer size range as in larger particles, so this would be an "existing" chemical. CNTs and fullerenes, however, even though also composed of carbon, do not exist with the same molecular structure as any "existing" chemical, and thus must be "new" chemicals. Unfortunately, the situation is much murkier than this. For example, do all CNTs have the same "molecular identity" and thus only have to be listed once, or are they all substantially different and thus each CNT manufacturer has to file a PMN? EPA addressed this specific question with a Notice published in the Federal Register on October 31, 2008 (EPA, 2008b), which read in part:

This document gives notice of the Toxic Substances Control Act (TSCA) requirements potentially applicable to carbon nanotubes (CNTs). EPA generally considers CNTs to be chemical substances distinct from graphite or other allotropes of carbon listed on the TSCA Inventory. Many CNTs may therefore be new chemicals under TSCA section 5. Manufacturers or importers of CNTs not on the TSCA Inventory must submit a premanufacture notice (PMN) (or applicable exemption) under TSCA section 5 where required under 40 CFR part 720 or part 723. In order to determine the TSCA Inventory status of a CNT, a manufacturer may submit to EPA a bona fide intent to manufacture or import under 40 CFR 720.25.

EPA has issued consent orders in response to several CNT PMNs. Typically, those orders have required 90-day inhalation toxicity testing, the use of engineering controls and personal protective equipment (PPE) during the manufacturing process, and restrictions on the final product (e.g., the CNT must be imbedded in a polymer or metal matrix in the commercial product).

In addition to the PMN provisions, TSCA also requires that persons who intend to manufacture, import, or process the chemical substances for an activity that is designated as a "significant new use" notify EPA at least 90 days before commencing that activity (EPA 2013a). This is known as the Significant New Use Rule (SNUR) and it applies to those who *use* covered materials, not just those who *produce* them. On February 25, 2013, EPA proposed SNURs for nanomaterials covered by previous PMN orders (EPA, 2013b):

> This proposed rule includes 14 PMN substances whose reported chemical names include the term "carbon nanotube" or "carbon nanofibers." Because of a lack of established nomenclature for carbon nanotubes, the TSCA Inventory names for carbon nanotubes are currently in generic form, e.g., carbon nanotube (CNT), multiwalled carbon nanotube (MWCNT), double-walled carbon nanotube (DWCNT), or single-walled carbon nanotube (SWCNT).

Because the SNURs are based on information submitted in PMNs and some of that information is confidential, EPA advises that:

> If an intended manufacturer, importer, or processor of CNTs is unsure of whether its CNTs are subject to this proposed SNUR or any other SNUR, the company can either contact EPA or obtain a written determination from EPA pursuant to the bona fide procedures at § 721.11

This discussion of the attempts to regulate CNTs under TSCA should convince the reader that the environmental regulation of ENMs in the United States is facing a long and tortuous path. As with occupational health regulations, manufacturers and users of ENMs and products containing them will have to move forward in an atmosphere of considerable uncertainty for many years to come. However, it is important to recognize the underlying rationale and intent of these laws. The TSCA PMN and SNUR requirements are intended to ensure that the hazards of new chemicals, or chemicals altered so that they are akin to new, be known to the authorities, so that the public can be protected. EPA's authority to order testing is a reflection of the idea that providers of materials should know whether they are providing new hazards along with their product. A company that investigates whether the materials they are producing or using are safe, and discloses what they learn, will appreciate the intent of these rules and will more likely avoid retroactive liabilities.

11.2.1.2 Other US Regulatory Programs In 2010, Tassinari and colleagues reviewed the status of environmental laws in the United States regarding nanotechnology (Tassinari et al., 2010) and concluded that TSCA was the only one where specific actions had taken place, either in EPA or in other agencies. With regard to the Consumer Product Safety Commission and the Occupational Safety and Health Administration, they state that each agency "considers its current policies sufficient for nanomaterials until more information is known." Their conclusion regarding the Food and Drug Administration is slightly different: "FDA considers its current practices sufficient to cover nanomaterials, but the agency will issue guidance on data to be included in submissions, including size."

Users of nanomaterials should be aware of other existing regulations and policies that may currently cover nanomaterials. These include:

- The Resource Conservation and Recovery Act (RCRA) is the nation's hazardous waste law. It requires proper identification and management of wastes that are either listed as hazardous, or which have certain characteristics that are defined as hazardous. If, by virtue of engineering particles at the nanoscale, new characteristics are imparted to a material that persist when it becomes waste material to be discarded, then the waste must be managed as a regulated hazardous waste. Some materials can become explosive or reactive dusts when manufactured on the nanoscale. Nanogold can be reactive, and nanoaluminum, when suspended in air at concentrations exceeding $30\,g/m^3$, can ignite at temperatures as low as $1000°C$ (Bouillard et al., 2010).

- The Clean Water Act requires that direct discharges to a water body do not substantially and adversely affect natural living things, and the EPA or an authorized state may require aquatic toxicity testing of any facility that is discharging a material that is causing mortality to aquatic life. Nanosilver and other materials have shown some indications of potential harm in this regard. A facility discharging to a water treatment facility, known as an indirect discharger, may have to meet additional requirements imposed by that facility, known as "local limits." These may be enforced in order to protect the microorganisms that function at the treatment facility to break down wastes.

- Public health agencies, which exist at the state and local level, have a general authority to protect public health and safety. They often respond to complaints with an order to desist the activity that causes complaints. Due to the fears that nanomaterials have raised, such agencies can be called into action when there is a perception of harm on the part of neighbors. For this reason, it is good practice to prevent air emissions of nanoparticles, for example, if a company is depending on ventilation to clear workspaces, it is advisable to collect the particles in filters as discussed in Section 10.1.2 and not just disperse them into the local neighborhood.

11.2.2 Environmental Regulations in the European Union

The EU is generally thought to have stronger chemical regulations than the United States. The current comprehensive chemical regulatory program on chemicals and their safe use is the law regarding Registration, Evaluation, Authorization and Restriction of

Chemical substances (REACH), which entered into force on June 1, 2007 (Commission of the European Communities, 2007). According to the EU:

> The aim of REACH is to improve the protection of human health and the environment through the better and earlier identification of the intrinsic properties of chemical substances. At the same time, REACH aims to enhance innovation and competitiveness of the EU chemicals industry. The benefits of the REACH system will come gradually, as more and more substances are phased into REACH.

Compared to TSCA, the REACH system places greater responsibility on industry to manage the risks from chemicals and to provide safety information on the substances. Manufacturers and importers are required to gather information on the properties of their chemical substances, which will allow their safe handling, and to register the information in a central database run by the European Chemicals Agency (ECHA) in Helsinki. The Agency acts as the central point in the REACH system: it manages the databases necessary to operate the system, coordinates the in-depth evaluation of suspicious chemicals, and is building up a public database in which consumers and professionals can find hazard information.

As with the EPA, a key question for nanomaterials is what constitutes a distinct chemical, or "substance," under REACH, since each such substance is subject to regulation. REACH uses a molecular/structural definition similar to that of EPA, discussed earlier. According to "Regulatory Aspects of Nanomaterials" (Commission of the European Communities, 2008a), "there are no provisions in REACH referring explicitly to nanomaterials. However, nanomaterials are covered by the 'substance' definition in REACH." Consequently, nanomaterials are required to comply with all REACH regulations, which are being phased in over a 11-year period. Some of the major provisions include:

- manufacturers and importers of at or above 1 metric ton/year of ENMs must submit a "registration dossier";
- if the quantity is at or above 10 metric tons/year, the manufacturer or importer must produce a chemical safety report;
- if deemed necessary, ECHA can require the providing of *any* information on an ENM;
- if a previously registered material is newly introduced in nanomaterial form, the registration dossier must be updated to provide information specific to the nanomaterial form; and
- the ECHA may require additional testing be performed on a nanomaterial, where "testing" is broadly defined.

Beyond the regulatory requirements of REACH, the EC has promulgated a recommended "code of conduct for responsible nanosciences and nanotechnologies (N&N) research" (Commission of the European Communities, 2008b). Key elements of the code of conduct include:

> N&N research activities should be safe, ethical and contribute to sustainable development serving the sustainability objectives of the Community as well as

contributing to the United Nations' Millennium Development Goals. They should not harm or create a biological, physical or moral threat to people, animals, plants or the environment, at present or in the future.

N&N research activities should be conducted in accordance with the precautionary principle, anticipating potential environmental, health and safety impacts of N&N outcomes and taking due precautions, proportional to the level of protection, while encouraging progress for the benefit of society and the environment.

Governance of N&N research activities should be guided by the principles of openness to all stakeholders, transparency and respect for the legitimate right of access to information. It should allow the participation in decision-making processes of all stakeholders involved in or concerned by N&N research activities.

The EC recommends:

That Member States encourage the voluntary adoption of the Code of Conduct by relevant national and regional authorities, employers and research funding bodies, researchers, and any individual or civil society organization involved or interested in N&N research and endeavor to undertake the necessary steps to ensure that they contribute to developing and maintaining a supportive research environment, conducive to the safe, ethical and effective development of the N&N potential.

11.3   COMPARISON OF NANOTECHNOLOGY REGULATION UNDER TSCA AND REACH

As discussed earlier, at this time, the two major regulatory frameworks being applied to nanotechnology are TSCA in the United States and REACH in the EU. Widmer and Meili (2010) have taken the instructive step to

...compare the two different regulatory approaches of REACH and TSCA in order to discuss some of the characteristics, issues and flaws that have been identified to be relevant in the process of governing manufactured nanomaterials, and possibly defining adaptations of the current regulatory frameworks in order that they are able to better handle the peculiarities of manufactured nanomaterials.

First, the authors point out that it may be fundamentally unfair to compare the two, since REACH took effect in 2007 and TSCA has been in force, essentially unchanged, since 1976. Nonetheless, they conclude that a comparison is warranted, since they are in fact the two regulatory schemes most relevant to ENMs. They then identify four issues of "paramount importance" for comparison: "the precautionary principle, the burden of proof, the differences in handling new and existing chemicals, and volume-based thresholds and exemptions." These issues are discussed in the following sections.

11.3.1   The Precautionary Principle and the Burden of Proof

The origins of the precautionary principle were discussed in Chapter 1. Widmer and Meili (2010) contend that closely related to the precautionary principle is the concept that the burden of proof—in this case, of harm from an ENM—"shifts" from a governmental entity (such as EPA or REACH) to the manufacturer of the material. REACH is explicitly based on the precautionary principle and a shifting of the burden of proof to the manufacturer. TSCA, on the other hand, does not mention the precautionary principle; rather, it requires that EPA demonstrate that a chemical "presents or will present an unreasonable risk" before the agency can take regulatory action.

On the other hand, both REACH and TSCA place the burden of proof for providing data to prove a new chemical's safety on its manufacturer. They differ, however, on how the data are then used in the decision to allow manufacturing. REACH takes the precautionary approach of requiring a manufacturer to prove that a chemical is safe before it is manufactured, while TSCA requires EPA to demonstrate that a chemical is harmful before it can prohibit its manufacture. It is clear that the EPA approach is the opposite of precautionary, since for a new material such as an ENM, it will undoubtedly take considerable time and effort to "prove" that it is unsafe. EPA can require that a company conduct toxicity testing, but in order to do so, it must demonstrate that data exist showing possible unreasonable risk of injury and potential for exposure. Widmer and Meili (2010) conclude:

> This evidentiary burden is regarded as one of TSCA's most significant flaws, particularly in regard to the handling of manufactured nanomaterials. The resulting situation is commonly referred to as a "catch-22"; the EPA must already have substantial information on a chemical in order to prove that it needs such information. These statutory requirements therefore essentially prevent the EPA from exercising its authority in cases when the scientific evidence is still fragmentary at best.

11.3.2 Differences in Handling New and Existing Chemicals

As discussed earlier, TSCA and REACH take similar approaches to defining different substances, based on their molecular structure. Neither piece of legislation explicitly distinguishes the nanometer-size version of a bulk material as a distinct, new material, although both are taking more tortuous paths to regulating nanomaterials. It is expected that both TSCA and REACH will have a growing list of ENMs that are considered to be "new" materials.

There are considerable differences between REACH and TSCA in the required actions for new and existing materials. Under TSCA, a company producing an ENM considered to be a new material must submit a PMN but, as discussed earlier, very little hard information is required as part of the submittal. This leads Widmer and Meili (2010) to the following conclusion.

> The EPA reportedly reviews about 1000 PMN every year, 67 per cent of which include no test data, and 85 per cent lack any health or ecotoxicity data. It seems therefore reasonable to conclude that the EPA will not receive the necessary data to perform sound risk assessment on manufactured nanomaterials from regular PMN submissions.

Once a material is defined as "existing," TSCA essentially considers it to be safe and has no requirements for data reporting and does not conduct further EHS assessments.

EPA does assess high-production volume chemicals (defined as a production of greater than 25,000 lb/year) under other initiatives, but most ENMs are not likely to reach this production volume for many years, if ever.

REACH takes a fundamentally different approach, by requiring that all chemicals whether new or existing meet the requirement outlined above. If a chemical has already been registered, new information must be submitted if a new use is identified or if new risk data are developed; presumably, both of these could apply to the development of a nanometer-sized version of an existing, registered chemical.

11.3.3 Volume-Based Thresholds and Exemptions

TSCA has several exemptions that may well apply to ENMs, including (Widmer and Meili, 2010):

- Low volume exemptions
- Certain categories of new materials are exempt from PMN review if they are produced or imported at less than 10 tons/year per manufacturer
- If an existing substance is produced or imported at less than 25,000 lb/year, it is exempt from providing updated information under the Inventory Update Rule (IUR)
- Low Release and Low Exposure substances are exempt from PMN reporting
- Small manufacturers are exempt from IUR reporting if they have either
  - Less than $40 million annual sales and less than 100,000 lb/year manufactured or imported or
  - Less than $4 million annual sales, regardless of quantity.

REACH also has increasing reporting requirements with increasing production volumes. Only chemicals with a production of 1 metric ton/year or less are exempt from reporting requirements, which is a considerably lower threshold than under TSCA.

11.4 PRIVATE LAW

It is always tempting, when evaluating the potential for legal liabilities, to narrow the universe that must be understood, so that the liabilities can appear to be manageable. Unfortunately, there is more to this than regulations. There is also the prospect of private suit. Anyone harmed by exposure to nanomaterials can sue in a court of law. Worker injury is generally covered by workers' compensation, and it is important to understand that the way to keep those costs down is to keep the rate of injury down. Beyond workers' compensation, the history of asbestos litigation has demonstrated that worker toxic tort litigation can wreak havoc on an industry that ignores worker health and safety.

Product liability is another matter. Makers and sellers of products can face suit for selling an inherently dangerous product (such as one that releases nanoparticles, causing exposures) that has either a manufacturing defect or a design defect or for selling a product with inadequate warnings. Although the potential for liability is very large, there are straightforward ways to manage these risks. These actions are consistent with what is recommended for complying with the intent of all of the environmental and health and safety rules above: make a good faith effort to understand the risks of using nanomaterials, make a good faith effort to prevent harm from them, and disclose what people would want to know. To understand what should be disclosed, simply ask what you would want to know if you were the purchaser, or the worker, who might be exposed to a hazard, and might wish to know how to avoid that exposure. It is when this simple, basic principle is ignored that liabilities result.

11.5   CONCLUSIONS

In Chapter 7, we argue that occupational and environmental exposure metrics based on mass have some shortcomings when it comes to nanoparticles, since a very high number concentration can have a very low mass. On the other hand, the alternatives, such as number or surface area, have significant practical limitations and their association with adverse health effects is still uncertain.

Likewise, the environmental regulation of ENMs using traditional mass-based exemptions and reporting rules raises difficult policy issues, since many ENMs are manufactured in very small mass quantities today, and likely will continue to do so for the foreseeable future. In Chapter 12, we suggest in general what form nanomaterial-specific regulations might possibly take to address these concerns.

REFERENCES

Bouillard J, Vignes A, Dufaud O, Perrin L, Thomas D. Ignition and explosion risks of nanopowders. J Hazard Mater 2010;181:873–880.

British Standards Institute [BSI] (2007) *Nanotechnologies—Part 2: Guide to Safe Handling and Disposal of Manufactured Nanomaterials*. British Standards Institute PD 6699-2:2007. London: BSI.

Commission of the European Communities. The introduction of measures to encourage improvements in the safety and health of workers at work. Brussels: Commission of the European Communities; 1989. Council Directive 89/391/EEC.

Commission of the European Communities. The protection of the health and safety of workers from the risks related to chemical agents at work. Brussels: Commission of the European Communities; 1998. Council Directive 98/24/EC.

Commission of the European Communities (2007) REACH. Available at http://ec.europa.eu/environment/chemicals/reach/reach_en.htm. Accessed November 4, 2014.

Commission of the European Communities (2008a) Communication from the Commission to the European Parliament, the Council and the European Economic and Social Committee—regulatory aspects of nanomaterials. COM(2008) 366 final.

Commission of the European Communities (2008b) Commission recommendation of 07/02/2008 on a code of conduct for responsible nanosciences and nanotechnologies research. COM(2008) 424 final.

Han JH, Lee EJ, Lee JH, So KP, Lee YH, Bae GN, Lee SB, Ji JH, Cho MH, Yu J. Monitoring multiwalled carbon nanotube exposure in carbon nanotube research facility. Inhal Toxicol 2008;20:741–749.

Myojo T, Oyabu T, Nishi K, Kadoya C, Tanaka I, Ono-Ogasawara M, Hirokazu Sakae H, Shirai T. Aerosol generation and measurement of multi-wall carbon nanotubes. J Nanopart Res 2009;11:91–99.

Oberdörster G, Stone V, Donaldson K. Toxicology of nanoparticles: a historical perspective. Nanotoxicology 2007;1:2–25.

Occupational Safety and Health Administration [OSHA]. Occupational Safety and Health Act of 1970. Washington (DC): U.S. Occupational Safety and Health Administration; 1970, Section 5.

Occupational Safety and Health Administration [OSHA]. Table Z-1 limits for air contaminants. Washington (DC): U.S. Occupational Safety and Health Administration; 2006. 29 CFR 1910:1000.

Occupational Safety and Health Administration [OSHA] (2007) Guidance for Hazard Determination for Compliance with the OSHA Hazard Communication Standard. U.S. Department of Labor, Occupational Safety and Health Administration. Available at https://www.osha.gov/dsg/hazcom/ghd053107.html. Accessed July 8, 2013.

Pauluhn J. Multi-walled carbon nanotubes (Baytubes): approach for derivation of occupational exposure limit. Regul Toxicol Pharmacol 2010;57:78–89.

Tassinari O, Bradley J, Holman M. The evolving nanotechnology environmental, health, and safety landscape: a business perspective. In: Hodge GA, Bowman DM, Maynard AD, editors. *International Handbook on Regulating Nanotechnologies*. Cheltenham: Edward Elgar; 2010. p 177–204.

Tsai SJ, Hofmann M, Hallock M, Ada E, Kong J, Ellenbecker MJ. Characterization and evaluation of nanoparticle release during the synthesis of single-walled and multi-walled carbon nanotubes by chemical vapor deposition. Environ Sci Technol 2009;43:6017–6023.

U.S. Environmental Protection Agency [EPA]. *TSCA Inventory Status of Nanoscale Substances—General Approach*. Washington (DC): EPA; 2008a.

U.S. Environmental Protection Agency [EPA]. Toxic Substance Control Act inventory status of carbon. U.S Environmental Protection Agency nanotubes. Washington (DC): U.S Environmental Protection Agency; 2008b. Federal Register, Volume 73, Number 212.

U.S. Environmental Protection Agency [EPA]. Nanoscale materials stewardship program interim report January 2009. Washington (DC): U.S Environmental Protection Agency; 2009.

U.S. Environmental Protection Agency [EPA] (2013a) TSCA Section 5 significant new use rules. Washington (DC): U.S Environmental Protection Agency. Available at http://www.epa.gov/opptintr/existingchemicals/pubs/sect5a2.html. Accessed July 10, 2013.

U.S. Environmental Protection Agency [EPA]. Proposed significant new use rules on certain chemical substances. Washington (DC): U.S Environmental Protection Agency; 2013b. 40 CFR Part 721.

U.S. National Institute for Occupational Safety and Health [NIOSH]. Diesel particulate matter (as elemental carbon). Washington (DC): U.S. National Institute for Occupational Safety and Health; 2003. Method 5040, Issue 3.

U.S. National Institute for Occupational Safety and Health [NIOSH]. Draft current intelligence bulletin—occupational exposure to carbon nanotubes and nanofibers. Washington (DC): U.S.

National Institute for Occupational Safety and Health; 2010. DHHS (NIOSH) Publication No. 2010–XXX

U.S. National Institute for Occupational Safety and Health [NIOSH]. Occupational exposure to titanium dioxide. Washington (DC): U.S. National Institute for Occupational Safety and Health; 2011. DHHS (NIOSH) Publication No. 2011–160.

U.S. National Institute for Occupational Safety and Health [NIOSH]. Current intelligence bulletin 65—occupational exposure to carbon nanotubes and nanofibers. Washington (DC): U.S. National Institute for Occupational Safety and Health; 2013. DHHS (NIOSH) Publication No. 2013–145.

van Broekhuizen P, Dorbeck-Jung B. Exposure limit values for nanomaterials—capacity and willingness of users to apply a precautionary approach. J Occup Environ Hyg 2013;10:46–53.

Widmer M, Meili C. Approaching the nanoregulation problem in chemicals legislation in the EU and US. In: Hodge GA, Bowman DM, Maynard AD, editors. *International Handbook on Regulating Nanotechnologies*. Cheltenham: Edward Elgar; 2010. p 238–267.

12

FUTURE DIRECTIONS IN ENGINEERED NANOPARTICLE HEALTH AND SAFETY

12.1 WHERE WE ARE TODAY

It is clear from the evidence discussed in Chapter 5 that there are serious concerns regarding the *toxicity* of some engineered nanomaterials (ENMs). It is also clear that work must be done to develop methods to *evaluate* exposures to ENMs (Chapters 7 and 8), *control* such exposures (Chapters 9 and 10), and *regulate* environmental and occupational exposures (Chapter 11). These needs have led to the development of a fairly robust nanoparticle health and safety research program in the United States, Europe, and elsewhere.

12.1.1 Research Efforts in the United States

In the United States, engineered naoparticles (ENP) research and policy is coordinated by the National Nanotechnology Initiative, or NNI, launched by the National Science and Technology Council, Committee on Technology, subcommittee on Nanoscale Science, Engineering, and Technology (NSTC, 2007). The purpose of the NNI is to coordinate nanotechnology activities among more than 25 federal agencies, including OSHA and NIOSH, the two federal agencies with primary responsibility for worker health and safety.

The general topic of environmental health and safety, including both occupational and environmental health, has been a major focus of the NNI. The NNI has four

Exposure Assessment and Safety Considerations for Working with Engineered Nanoparticles,
First Edition. Michael J. Ellenbecker and Candace Su-Jung Tsai.

goals; the first three deal with technology development and education and the fourth goal is "support responsible development of nanotechnology" (NSTC, 2007):

> The NNI aims to maximize the benefits of nanotechnology and at the same time to develop an understanding of potential risks and to develop the means to manage them. Specifically, the NNI pursues a program of research, education and communication focused on environmental, health, safety, and broader societal dimensions of nanotechnology development.

In a recent document (NSTC, 2011), the NNI has described its EHS research initiatives in some detail. They fall into six broad categories: measurement methods, human exposure assessment, human health, the environment, risk assessment and risk management methods, and informatics and modeling. All these, except the environment, are relevant to the issues of worker health and safety discussed in this book. The goals of the research in each area address many of the issues that we have discussed.

Under measurement methods, the research goals are:

- Develop measurement tools to detect and identify engineered nanoscale materials in products and relevant matrices and determine their physicochemical properties throughout all stages of their life cycles.
- Develop measurement tools for determination of biological response and enable assessment of hazards and exposure for humans and the environment from ENMs and nanotechnology-based products throughout all stages of their life cycles.

Under these broad goals, the following research needs were identified, along with (in parentheses) the number of funded projects and the funding allocated to them by federal agencies in fiscal year 2009:

1. Develop measurement tools for determination of physicochemical properties of ENMs in relevant media and during the life cycles of ENMs and (nanotechnology-enabled products (NEPs) (18, \$4.8 M).
2. Develop measurement tools for detection and monitoring of ENMs in realistic exposure media and conditions during the life cycles of ENMs and NEPs (14, \$4.3 M).
3. Develop measurement tools for evaluation of transformations of ENMs in relevant media and during the life cycles of ENMs and NEPs (6, \$1.2 M).
4. Develop measurement tools for evaluation of biological responses to ENMs and NEPs in relevant media and during the life cycles of ENMs and NEPs (2, \$0.6 M).
5. Develop measurement tools for evaluation of release mechanisms of ENMs from NEPs in relevant media and during the life cycles of NEPs (\$0).
6. Projects that benefit the research category (2, \$400 K).

In the human exposure assessment category, the following goals were identified:

- Identify potential sources, characterize the exposure scenarios, and quantify actual exposures of workers, the general public, and consumers to nanomaterials.
- Characterize and identify the health outcomes among exposed populations in conjunction with information about the control strategies used and exposures to determine practices that result in safe levels of exposures.

Under these goals, the following research needs (and projects and expenditures in FY 09) were identified:

1. Understand processes and factors that determine exposures to nanomaterials (5, $900 K).
2. Identify populations groups exposed to ENMs (1, $1.1 M).
3. Characterize individual exposures to nanomaterials (7, $1.2 M).
4. Conduct health surveillance of exposed populations (1, $45 K).

In the human health category, the following goals were identified:

- Understand the relationship of physicochemical properties of engineered nanoscale materials to *in vivo* physicochemical properties and biological response.
- Develop high-confidence predictive models on *in vivo* biological responses and causal physicochemical properties of ENMs.

Under these goals, the following research needs (and projects and expenditures in FY 09) were identified:

1. Identify or develop appropriate, reliable, and reproducible *in vitro* and *in vivo* assays and models to predict *in vivo* human responses to ENMs (17, $5.2 M).
2. Quantify and characterize ENMs in exposure matrices and biological matrices (19, $7.9 M).
3. Understand the relationship between the physicochemical properties of ENMs and their transport, distribution, metabolism, excretion, and body burden in the human body (7, $2.7 M).
4. Understand the relationship between the physicochemical properties of ENMs and uptake through the human port-of-entry tissues (13, $3.8 M).
5. Determine the modes of action underlying the human biological response to ENMs at the molecular, cellular, tissue, organ, and whole-body levels (40, $17.9 M).
6. Determine the extent to which life stage and/or susceptibility factors modulate health effects associated with exposure to ENMs and NEPs and applications (3, $1.3 M).
7. Projects that meet multiple goals and projects that benefit the research category (18, $2.8 M).

In the environment category, the following goal was identified:

- Understand the environmental fate, exposure, and ecological effects of ENMs, with priority placed on materials with highest potential for release, exposure, and/or hazard to the environment.

Under this goal, the following research needs (and projects and expenditures in FY 09) were identified:

1. Understand environmental exposures through the identification of principal sources of exposure and exposure routes (2, $250 K).
2. Determine factors affecting the environmental transport of nanomaterials (4, $1.3 M).
3. Understand the transformation of nanomaterials under different environmental conditions (16, $5.1 M).
4. Understand the effects of ENMs on individuals of a species and the applicability of testing schemes to measure effects (4, $715 K).
5. Evaluate the effects of ENMs at the population, community, and ecosystem levels (1, $0).
6. Projects that capture multiple needs (27, $36.3 M).

In the risk assessment and risk management category, the following goal was identified:

- Increase available information for better decision making in assessing and managing potential risks from nanomaterials, including using comparative risk assessment and decision analysis; life cycle considerations; and additional perspectives such as ELSI (ethical, legal and social issues) considerations, stakeholders' values, and additional decision makers' considerations.

Under this goal, the following research needs (and projects and expenditures in FY 09) were identified:

1. Incorporate relevant risk characterization information, hazard identification, exposure science, and risk modeling and methods into the safety evaluation of nanomaterials (6, $1.8 M).
2. Understand, characterize, and control workplace exposures to nanomaterials (10, $1.3 M).
3. Integrate life cycle considerations into risk assessment (4, $300 K).
4. Integrate risk assessment into decision-making frameworks for risk management ($0).
5. Integrate and standardize risk communication within the risk management framework (1, $200 K).

The final research category identified in the NNI research strategy is "informatics and modeling for nanoEHS research." The discussion of this category indicated the need for research on such topics as data acquisition, data analysis, data sharing, structural modeling of ENMs, predictive models and simulations, and the building of an informatics infrastructure. No specific goals nor project money was identified in this category.

A little quick math reveals that a total of 248 EH&S research projects worth $104 million were funded in FY 09. While this might appear to be a substantial government investment into health and safety, it needs to be compared to the total federal investment into nanoparticle research, which was $1.7 billion in FY 09. Prominent nanoparticle EHS researchers have called for a substantial increase in federal funding for this field. For example, in 2007 testimony before the US House of Representatives Committee on Science and Technology, Andrew Maynard, the then Chief Science Advisor, Project on Emerging Technologies at the Woodrow Wilson International Center for Scholars and now the Director of the Center for Risk analysis at the University of Michigan, recommended that at least 10% of the federal nanotechnology research budget be dedicated to "strategic risk-related research" (Committee on Science and Technology, U.S. House of Representatives, 2007). This would represent a significant increase over present funding levels.

12.1.2   Research Efforts in Europe

In response to the regulatory need to protect the health of those working with ENPs, the EU has embarked on an extensive research program encompassing toxicology and the development of techniques to evaluate and control exposures. The EU defines their funded research programs in terms of "framework programs"; the sixth framework program, covering the 2002–2006 time period, and the seventh framework, covering 2007–2013, included funding for research in this area.

12.1.3   Progress toward Research Goals

As important as the question of the *level* of research funding is the question of the *topics* addressed by the research. Do the research projects detailed above answer the right questions? Andrew Maynard and coauthors took a broad look at this question in a commentary published in *Nature* in 2006 (Maynard et al., 2006), where they laid out five "grand challenges" that must be addressed to ensure the development of sustainable nanotechnology. It is instructive to look at these challenges and assess the progress that has been made in each area over the ensuing 7 years.

12.1.3.1   Challenge 1. Develop Instruments to Assess Exposure to Engineered Nanomaterials in Air and Water, within the Next 3–10 Years   As discussed in Section 7.2, considerable progress has been made in developing instruments for evaluating airborne nanoparticles, and this has been an area of significant funding under the first NNI research area, measurement. However, as discussed in Section 12.3,

available instruments remain expensive, bulky, and difficult to use. Maynard et al. (2006) addressed this specific issue:

> But people working with nanomaterials urgently need inexpensive personal aerosol samplers that are capable of measuring exposure in the workplace and environment. This universal aerosol sampler would log exposure against aerosol number, surface area and mass concentration simultaneously, and provide a historic record that can be interpreted in the light of new knowledge and new exposure monitoring paradigms. It would be portable, sufficiently inexpensive to ensure widespread use, and available commercially within the next 3 years.

This ambitious goal was certainly not met by 2013; we are no closer to having this ideal instrument when this book was written in 2014 than we were in 2006.

Regarding nanoparticles in water, the authors state that the need "…is to develop instruments that can track the release, concentration and transformation of ENMs in water systems (including liquid-based nanotechnology consumer products), within the next 5 years."

Finally, the authors call for the development of "…smart sensors that indicate potential harm to human health. An example would be sensors that simultaneously detect airborne nanoparticles and determine their potential to generate reactive oxygen species—possibly providing early indications of harm. Such sensors should be available within the next 10 years." Researchers have made some progress toward this goal, but practical devices are likely still many years away.

12.1.3.2 Challenge 2. Develop and Validate Methods to Evaluate the Toxicity of Engineered Nanomaterials, within the Next 5–15 Years

It is fair to say that nanoparticle toxicology has received the most nano-EHS research funding since 2006; as detailed above, toxicology studies constituted more than 40% of the NNI EHS budget in FY 09. Maynard et al. laid out some specific goals under this grand challenge, that is, "validated screening tests, developing viable alternatives to *in vivo* tests, and determining the toxicity of fibre-shaped nanoparticles." With regard to validated screening tests, they propose that the nanoparticle EHS community "…reach international agreement on a battery of *in vitro* screening tests for human and environmental toxicity within the next 2 years, and to validate these tests within the next 5 years." While a great deal of progress has been made toward developing screening tests, as described in Chapter 5, we are still a long way from reaching this ambitious goal.

Closely related to *in vitro* screening tests is the next goal, developing alternatives to *in vivo* toxicity testing.

Considerable progress has been made toward the third toxicology goal regarding fiber-shaped nanoparticles, especially carbon nanotubes (CNTs) and to some extent carbon nanofibers (see Section 5.4). Researchers are developing other nanoparticles with large aspect ratios, such as nanowires being incorporated into nanoscale electronic devices; little is known about the toxicity of these particles.

12.1.3.3 Challenge 3. Develop Models for Predicting the Potential Impact of Engineered Nanomaterials on the Environment and Human Health, within the Next 10 Years This broad challenge also has three parts as Maynard et al. (2006) states:

> First, to develop validated models capable of predicting the release, transport, trans-formation, accumulation and uptake of engineered nanomaterials in the environment. In parallel, validated models must be developed that are capable of predicting the behaviour of engineered nanomaterials in the body, including dose, transport, clearance, accumulation, transformation and response ... Third, to use predictive models for engineering nanomaterials that are safe by design.

The first part of this challenge is relevant to this discussion, and it appears that progress has been slow on the development of such models. We are still early in the process of collecting basic data that can be used in the development of such models.

12.1.3.4 Challenge 4. Develop Robust Systems for Evaluating the Health and Environmental Impact of Engineered Nanomaterials Over Their Entire Life, Within the Next 5 Years This is a call to develop life cycle assessment methods applicable to ENMs. As discussed in Section 6.5, little progress has been made to date on any phase of a nanomaterial life cycle beyond manufacturing.

12.1.3.5 Challenge 5. Develop Strategic Programs that Enable Relevant Risk-Focused Research, Within the Next 12 Months This was a short-term goal, presented in 2006, to develop the research agenda described above. Some of their specific recommendations are noteworthy. First, the authors call for the development of "...collaborative research programmes—whether interdisciplinary, between government and industry, or between different stakeholders." This call has been repeated in international meetings on nanoparticle health and safety, but in our opinion little progress has been made. For example, the broad field of toxicology seems to be characterized by strong competition between research groups, which can lead to excellent results but can also result in the duplication of efforts. In exposure assessment, efforts are under way as of this writing to develop international consensus protocols for evaluating exposure. For example, an informal international group of scientists concerned with ENP exposure assessment has met several times, with the goal of developing consensus exposure assessment proto-cols for different types of ENPs.

Maynard et al. also recommend that attention be paid to the communication of research results to interested and affected non-scientific audiences, including government personnel and the general public. In response to this need, it is fair to say that the number of publications, scientific symposia, intergovernmental meetings, and so on, has increased dramatically in recent years. Communicating with the general public regarding the risks and benefits of nanotechnology is much more difficult than communication between scientists and government regulators; Berube and colleagues (2010) present an excellent discussion regarding this issue. The authors finish with this recommendation:

> Finally, a global understanding of nanotechnology-specific risks is essential if large and small industries are to operate on a level playing field, and developing economies are not to be denied essential information on designing safe nanotechnologies. We propose that mechanisms, networks and meetings are established that enable international information-sharing and coordination between the public and private sectors.

We can only share in the desire for an ambitious program to accomplish this goal.

12.2 HUMAN HEALTH EFFECTS STUDIES

As discussed in Chapter 5, all of our concerns regarding the possible health effects associated with ENMs arise from animal studies. However, it is generally accepted in the scientific community that the association between an exposure to any toxic material and a particular health effect can only be confirmed by epidemiologic studies. For example, the International Agency for Research on Cancer (IARC) considers an agent to be a "confirmed human carcinogen" (Group 1) when "…there is sufficient evidence of carcinogenicity in humans." Generally, this requires evidence from high-quality epidemiologic studies. If excess rates of a carcinogen have been found in animal studies along with limited evidence for carcinogenicity in human populations, the agent is considered to be a "probable human carcinogen" (Group 2A); limited evidence for carcinogenicity in humans combined with "…less than sufficient evidence of carcinogenicity in experimental animals" results in Group 2B, a "possible human carcinogen" (IARC, 2006). Either "sufficient" or "limited" evidence in humans requires positive epidemiologic studies.

As of this writing, there have been no published epidemiology studies for any ENM; this lack is due to issues in both exposure and response. One condition limiting the feasibility of conducting epidemiologic studies is the limited size of any exposed population; the nanotechnology industry is still in its relative infancy and consequently it has not been possible to assemble a large enough cohort of possibly exposed workers. Concerning response, the infancy of the industry probably limits any epidemiology study to acute health effects; the issue is to identify the most important health end points that might be associated with an acute response to ENMs.

12.3 EXPOSURE ASSESSMENT

12.3.1 Future Needs in Exposure Assessment Techniques

In occupational hygiene, exposure assessment studies can be performed for several reasons, but the two most important can be categorized broadly as exposure assessment for research and exposure assessment to ensure that workers are not being overexposed. In the first case, exposure assessment may be performed for example as part of an epidemiologic study, where the goal is to investigate possible associations between exposure and disease. In the second case, routine monitoring might be performed by a plant occupational hygienist to ensure that worker exposures are below any applicable occupational exposure limits (OELs).

For both exposure assessment categories, it is very desirable that standardized equipment and methodologies be used. For example, the term "respirable mass sampling" means the same thing to occupational hygienists throughout the world, since standardized instruments and methodologies have been developed and promulgated by government agencies and professional organizations such as the American Conference of Governmental Hygienists (ACGIH, 2012).

For research, standardization allows the comparison of studies performed by different research groups, and perhaps the pooling of data from several studies to allow what is called a meta-analysis of those data. For routine monitoring, implicit (and sometimes explicit) in any OEL is the exposure assessment *method* to be used in evaluating exposures for comparison to that OEL.

As discussed in Chapter 7, the equipment that is currently used to evaluate exposures to ENPs can be characterized as expensive, bulky, and difficult to use. These instruments are best suited for research studies and are unsuitable for routine monitoring. While improved research instruments remain a high priority, the crucial need at this time is for instruments that can be used by plant occupational hygienists for routine monitoring. To meet this need, the instruments must be relatively inexpensive, battery-operated and portable, and relatively easy to use. It is also highly desirable that equipment be developed that can be worn by workers, so that personal breathing zone (BZ) samples can be collected. And, of course, the instruments must be capable of measuring the proper end point for comparison to the applicable OEL; this issue is discussed further in Section 12.3.2.

In the research community, there is a great deal of discussion at this time as to the proper methodology to be used to evaluate exposure. There is a general recognition of the need to develop consistent exposure assessment techniques to be used by research groups in all countries. The EU, because of its existing structure and funding methods, has made the most progress toward this goal. Funded research programs include:

- NANODEVICE, whose objective is "to develop innovative concepts and reliable methods for characterizing ENP in workplace air with novel, portable and easy-to-use devices suitable for workplaces (EU, 2013a)."
- NanoGEM, a project funded by Germany, addresses a range of issues relative to sustainable development and risk assessment for ENMs, including exposure assessment (BMBF, 2013). NANODEVICE and NanoGEM have funded the development of a Nano Test Center which can generate standardized test aerosols for the intercomparison of NP instruments.
- SANOWORK, whose objective is "...to identify a safe occupational exposure scenario by exposure assessment in real conditions and at all stages of nanomaterials (NM) production, use and disposal. In order to address this ... we intend

to … implement a rigorous exposure assessment in the workplace in order to evaluate the effectiveness of existing and proposed exposure reduction strategies" (EU, 2013b).

Actual progress toward this goal beyond the EU, however, has been hindered by structural and financial barriers. Since there is no worldwide agency that is attempting to coordinate this effort, it is being carried forward largely by informal efforts of leading nanoparticle health and safety scientists, who have formed the Global Measurement Harmonization Workgroup. This group has had several meetings; as of this writing, considerable progress has been made toward the development of consensus exposure assessment methods.

12.3.2   The Development of Occupational Exposure Limits

As of this writing, very little progress has been made toward the development of OELs, although the topic has been the subject of much discussion among professionals in the field (Gordon et al., 2014). In the United States, NIOSH has published recommended exposure limits (RELs) for nanometer-sized TiO_2 and CNTs (see Section 11.1). As discussed earlier, these are both mass-based standards, which are problematic in a number of ways. An alternative approach, using particle number concentration, would overcome many of the shortcomings of mass-based standards. At this time, there is only one number concentration personal exposure limit and threshold limit value, that is, for asbestos. Asbestos OELs typically specify the allowable number concentration, the sizes of fibers to be counted, and the measurement method. This illustrates some of the difficulty in setting any number-based OELs for ENMs—any standard must be based on a measurement method.

In general, the equipment used to measure exposure for comparison to an OEL must meet the following requirements:

- It must be able to measure the specific substance covered by the OEL;
- It must be wearable by the worker, so that a personal BZ sample can be collected;
- It must be able to measure concentrations lower than the OEL;
- It must be straightforward and fairly simple to operate, so that field occupational hygienists with minimal training can successfully operate it; and
- Both the equipment used to collect the BZ sample and the analytical method used to quantify it should be relatively inexpensive, so that a wide range of companies and government agencies can afford to collect samples.

Unfortunately, the equipment currently available to measure particle number concentration, as described in Section 7.2, meets none of these requirements, except possibly the third one. The equipment used to collect the data presented in the various case studies in previous chapters is heavy, expensive, and difficult to operate.

The analysis of collected samples in order to identify specific ENPs is also expensive and time-consuming. This is an obvious disincentive to developing number-based OELs and is one of the primary reasons that the NIOSH TiO_2 and CNT RELs propose mass-based standards. There is a great need for instrument manufacturers to develop instruments that meet the above requirements, in order that adequate OELs can be adapted for protecting workers.

12.4   OPTIMAL APPROACHES TO CONTROL EXPOSURES

As discussed in Chapters 9 and 10, the occupational and environmental health and safety community's knowledge concerning the control of occupational and environmental exposures is much further developed than our knowledge of health effects and exposure assessment. Given that, there is still a great deal of work to be done in this area before we can be fairly confident that workers are not being exposed to dangerous levels of ENMs and that ENMs are not being released to the environment at levels that pose a risk to humans and other species. Important issues are discussed in the following.

12.4.1   Engineering Control of Occupational Exposures

The good news regarding engineering controls is that the standard methods used by occupational hygienists for other exposures can be used to control exposures to ENPs. The bad news is that those controls may be required to have extremely high control efficiencies, since even small releases of ENPs may be of concern. As discussed in Chapter 9, a typical laboratory fume hood may appear to be effectively controlling ENPs being handled in the hood, when in fact a significant number of those particles can be reaching the worker's BZ.

Until we know much more about the toxicity of different ENPs, it is imperative that we design and use control systems that reduce exposures to the lowest possible level. The basic lesson learned from our research on control systems is that their design must be optimized for use with ENPs, and their performance must be carefully evaluated under actual use conditions to ensure that exposures are being reduced as close as possible to background levels. It would be a serious mistake to assume that the standard hood designs for specific operations published in the *Ventilation Manual* (ACGIH, 2007) and other reference books will be adequate for controlling ENPs.

For ENPs in dry powder form, the optimum control strategy in most situations will be a combination of isolation and local exhaust ventilation. Much more research is needed on the optimization of control systems for specific applications involving the handling of dry powder ENMs. For straightforward tasks such as the weighing or transfer of small quantities of dry powder ENPs, our research has shown that the specially designed powder handling enclosures discussed in Section 8.1.2 are very effective at minimizing nanoparticle release and exposure.

12.4.2 Control Banding

Control banding is

> ...a qualitative strategy for assessing and managing hazards associated with chemical exposures in the workplace. The concept is used to manage exposures to potentially hazardous materials through the application of one of four recommended control approaches. The concept is based on the premise that although many chemical hazards exist, there are a limited number of controls available. To determine the appropriate control strategy, one must consider the characteristics of the substance, the potential for exposure, and the hazard associated with the substance. As the potential for harm to the worker increases, so does the degree of control needed to manage the risk. (NIOSH, 2012)

The four control bands range from requiring minimal control to complete emission control:

1. Band 1: use general ventilation and good occupational hygiene practice.
2. Band 2: use local exhaust ventilation.
3. Band 3: completely enclose the process.
4. Band 4: consult with experts to develop complete control.

Although several control banding tools have been developed for ENMs, there is little evidence in the literature of the efficacy of this approach for this specific application. More research is needed on the application of this approach, and whether the resulting exposures are considered to be satisfactorily controlled.

12.4.3 Respiratory Protection

If measurements indicate that workers are being exposed to elevated concentrations of ENPs, or if, in the absence of measurements, sufficient concern exists as to *possible* ENP exposures, respiratory protection should be used as an interim control measure until engineering controls can be instituted to reduce exposures to an acceptable level. As discussed in Section 9.4, research has demonstrated that NIOSH-certified respirator cartridges perform to their specifications for ENPs, just as they do for larger particles—high efficiency particulate air filter (HEPA)-equipped P/N100 cartridges collect close to 100% of nanoparticles of all sizes and P/N95 cartridges collect at least 95% of particles across the nanoparticle size range.

As also discussed in Section 9.4, respirators do not offer complete protection because of the potential for face seal leakage. Research performed to date has only evaluated the performance of respirator cartridges against ENPs and not the performance of entire respirators worn by workers under actual or simulated working conditions. In particular, the penetration of ENPs by face seal leakage has not been studied. Such research is badly needed before we can have full confidence in the protection afforded by air-purifying respirators.

12.4.4 Safe Work Practices

Many documents describe safe work practices to be followed when working with ENMs; the most recent, comprehensive document is that produced jointly by NIOSH and the authors of this book through the Center for High-rate Nanomanufacturing (CHN) titled "General Safe Practices for Working with Engineered Nanomaterials in Research Laboratories" (NIOSH, 2012). This topic is well understood and does not require innovative future research.

12.4.5 Air Pollution Control of Nanoparticles

As discussed in Chapter 9, the only air pollution control technique that is known to be effective for airborne nanoparticles is HEPA filtration, which has limited application in manufacturing due to the high cost associated with the inability to clean the filters. Of the other major categories available for controlling airborne particles, the options are limited. Electrostatic precipitation is not likely to be effective, due to the difficulties in charging particles in the nanoparticle size range. Venturi scrubbers may be effective at collecting airborne nanoparticles, but this device type has the inherent disadvantage that it trades an air pollution problem for a water pollution problem.

That leaves fabric filtration, which is likely to be an effective control technique, since the filtration collection efficiency of a typical filter is at a minimum for particle diameters of several hundred nanometers and increases as the particle size decreases from that range (see Section 10.1). There is need for research into fabric filter performance for nanoparticles, however, as the published literature has focused on larger particles. As discussed in Section 10.6, we have recently published a pilot study of the collection efficiency of a range of fabric filter fabrics when challenged with ENPs (Tsai et al., 2012), which found that certain fabrics can effectively collect ENPs. This is consistent with previous research on larger particles, which indicates that fabric filters generally have high collection efficiency, but they typically allow the penetration of a small fraction of the aerosol, usually associated with the cleaning mechanism. Thus, a likely approach for the effective control of nanoparticles at very high efficiency is a two-stage device consisting of a fabric filter followed by a HEPA filter. The fabric filter will remove the bulk of the aerosol, and the HEPA filter will deliver very low particle penetration and will have a long life time due to the low particle concentration escaping from the fabric filter.

12.4.6 Water Pollution Control of Nanoparticles

As discussed in Section 10.2, a great deal of research is needed into the control of waterborne ENPs by sewage treatment plants. Research is needed both on the efficiency of ENP removal by sewage treatment and the possible adverse effects of ENMs on the overall performance of such plants.

12.4.7 Nanoparticles in Waste Streams

Section 10.3 described the limited information available at this time concerning the fate and transport of ENPs once they enter the waste stream. A great deal of research is needed to further characterize the processes that affect such particles. Hasselov and Kaegi (2009) give the following exhaustive list of factors that require further research:

- Dispersion stability in water
- Biotic degradability
- Ready biodegradability
- Sewage treatment simulation testing
- Identification of degradation product(s)
- Further testing of degradation product(s) as required
- Abiotic degradability and fate
- Hydrolysis, for surface-modified materials
- Adsorption–desorption
- Adsorption to soil or sediment
- Bioaccumulation potential.

12.5 THE FUTURE OF REGULATION

As discussed in Chapter 11, with respect to ENMs the current status of regulations in the United States can kindly be described as inadequate. Since history tells us that US regulations respond slowly to emerging materials, it is likely that it will be quite some time before existing environmental and occupational health regulations directly address ENMs. The fundamental problem, as many have noted (Geiser, 2001), is structural; the US approach, as embodied in the major environmental and occupational health laws of the 1970s, is proscriptive, setting emission limits on a substance-by-substance and medium-by-medium basis. On the other hand, the EU approach, as embodied in REACH, is comprehensive and precautionary. It is very likely that occupational and environmental health regulations specific to nanomaterials will first be developed in the EU.

12.6 CONCLUSIONS

With nanotechnology, we are at a time that is both exciting and fraught with danger. The benefits to society from this new field are likely to be great, but the risks to human health and the environment may also be very large. Luckily, we have the ability to take advantages of the benefits while minimizing the risks. As we learn more about the actual magnitude of those risks, we can take the precautionary approach of minimizing ENM exposures and releases to the environment. The technical means to do this are known, and the costs will be manageable.

This precautionary approach must begin now, and it must be accomplished largely in the absence of regulation. This means that all of the affected parties—manufacturers of ENMs and ENM-enabled products, governments, labor, the health and safety community, and the public—must join together in a proactive, cooperative effort to protect workers, the public, and the environment.

We will end where we began, with asbestos. In 2000, almost a century after the adverse health effects of asbestos exposure were first found, an estimated 43,000 people worldwide died from mesothelioma and many more from lung cancer and other asbestos-related diseases (LaDou et al., 2010). It is too late to protect our parents and grandparents from that deadly scourge, but it is not too late to ensure that our children will not face a similar fate from ENMs.

REFERENCES

American Conference of Governmental Industrial Hygienists [ACGIH]. *Industrial Ventilation: A Manual of Recommended Practice for Design*. Cincinnati (OH): ACGIH; 2007.

American Conference of Governmental Industrial Hygienists [ACGIH]. *2012 TLVs and BEIs—Based on the Documentation of the Threshold Limit Values for Chemical Substances and Physical Agents & Biological Exposure Indices*. Cincinnati (OH): American Conference of Governmental Industrial Hygienists; 2012.

Berube DM, Faber B, Scheufele DA. Communicating risk in the 21st Century: the case of nanotechnology. Public Communication of Science and Technology Project. Raleigh (NC): North Carolina State University; 2010.

Bundesministerium fur Bildung und Forschung [BMBF]. 2013. NanoGEM. BMBF (Federal Ministry of Education and Research). Available at http://www.nanopartikel.info/en/projects/completed-projects/nanogem. Accessed March 21, 2013.

Committee on Science and Technology, U.S. House of Representatives. Research on environmental and safety impacts of nanotechnology: current status of planning and implementation under the National Nanotechnology Initiative. Washington (DC): U/S Government Printing Office; 2007. Serial No. 110-69.

European Union [EU]. 2013a. NANODEVICE. European Union. Available at http://www.nano-device.eu/index.php?id=249. Accessed March 19, 2013.

European Union [EU]. 2013b. Sanowork. European Union. Available at http://www.sanowork.eu/. Accessed March 19, 2013.

Geiser K. *Materials Matter—Toward a Sustainable Materials Policy*. Cambridge (MA): MIT Press; 2001.

Gordon SC, Butala JH, Carter JM, Elder A, Gordon T, Gray G, Sayre PG, Schulte PA, Tsai CS, West J. Workshop report: strategies for setting occupational exposure limits for engineered nanomaterials. Regul Toxicol Pharmacol 2014;68:305–311.

Hassellov M, Kaegi R. Analysis and characterization of manufactured nanoparticles in aquatic environments. In: Lead JR, Smith E, editors. *Environmental and Human Health Impacts of Nanotechnology*. Chichester: John Wiley & Sons, Ltd.; 2009. p 253.

International Agency for Research on Cancer [IARC]. *Preamble, IARC Monographs on the Evaluation of Carcinogenic Risk to Humans*. Lyons: IARC; 2006.

LaDou J, Castleman B, Frank A, Gochfeld M, Greenberg M, Huff J, Joshi TK, Landrigan PJ, Lemen R, Myers J, Soffritti M, Soskolne CL, Takahashi K, Teitelbaum D, Benedetto Terracini B, Watterson A. The case for a global ban on asbestos. Environ Health Perspect 2010;118:897–901.

Maynard A, Aitken RJ, Butz T, Colvin V, Donaldson K, Oberdorster G, Philbert MA, Ryan J, Seaton A, Stone V, Tinkle S, Tran L, Walker N, Warheit DB. Safe handling of nanotechnology. Nature 2006;444:267–269.

National Science and Technology Council [NSTC]. *The National Nanotechnology Initiative—Strategic Plan*. Washington, DC: NSTC, the National Nanotechnology Initiative, Executive Office of the President of the United States; 2007.

National Science and Technology Council [NSTC]. National Nanotechnology Initiative—Environmental, Health, and Safety Research Strategy. Washington (DC): NSTC, the National Nanotechnology Initiative, Executive Office of the President of the United States; 2011.

Tsai SJ, Echevarría-Vega M, Sotiriou G, Santeufemio C, Huang C, Schmidt D, Demokritou P, Ellenbecker M. Evaluation of environmental filtration control of engineered nanoparticles using the Harvard Versatile Engineered Nanomaterial Generation System (VENGES). J Nanopart Res 2012;14:812.

U.S. National Institute for Occupational Safety and Health [NIOSH]. *General Safe Practices for Working with Engineered Nanomaterials in Research Laboratories*. Cincinnati (OH): NIOSH; 2012.

INDEX

Exposure Assessment and Safety Considerations for Working with Engineered Nanoparticles,
First Edition. Michael J. Ellenbecker and Candace Su-Jung Tsai.
© 2015 John Wiley & Sons, Inc. Published 2015 by John Wiley & Sons, Inc.